Information Science and Statistics

Series Editors:
M. Jordan
J. Kleinberg
B. Schölkopf

T0185160

Information Science and Statistics

Jorma Rissanen

Information and Complexity in Statistical Modeling

Springer

Jorma Rissanen
Helsinki Institute for Information
   Technology
Helsinki, Finland
jrissanen@yahoo.com

Series Editors:

Michael Jordan
Division of Computer
   Science and
   Department of
Statistics
University of California,
   Berkeley
Berkeley, CA 94720
USA

Jon Kleinberg
Department of Computer
   Science
Cornell University
Ithaca, NY 14853
USA

Bernhard Schölkopf
Max Planck Institute for
   Biological Cybernetics
Spemannstrasse 38
72076 Tübingen
Germany

ISBN 978-1-4419-2267-0
eISBN-10: 0-387-68812-0
eISBN-13: 978-0-387-68812-1

Printed on acid-free paper.

9 8 7 6 5 4 3 2 1

springer.com

Contents

Preface

Shannon's formal measure of information, the entropy together with the related notions of relative entropy or the Kullback–Leibler distance, the mutual information, and the channel capacity, are all defined in the mean sense with respect to a given probability distribution. They were meant for use in communication problems, where the statistical properties of the channels through which the messages are transmitted are quite well estimated before the messages are sent. Also the word *information* actually refers to the number of messages as typical samples from the various probability distributions. The greater the number of the typical messages is, the greater the information; and, for instance, the entropy amounts to the per symbol limit of the logarithm of the number of typical messages, which is hardly what we mean by *information* in everyday language.

In statistical modeling there is only one data set from which the statistical properties of the data constituting the model are to be learned, and there are hoped to hold for future data generated by the same physical process. This time the word *information* means the amount of these learnable properties, and Shannon's information would be *complexity* as the code length needed to encode the data. The confusion regarding the meaning of the word *information* together with the different nature of the basic problems appear to be the reason why Shannon's elegant measures have never found extensive applications in statistical problems despite several attempts by probability theorists and statisticians. And yet coding theory, which most statisticians are not familiar with, is absolutely fundamental in statistical applications for the simple reason that code length is in one-to-one correspondence with probability measurements.

In the mid-1960s a remarkable notion of information of an individual object, based on the concepts of computation rather than probabilities, was introduced independently by Solomonoff, Kolmogorov, and Chaitin – in that chronological order: namely, the information in a binary string, or its *complexity* as it was called by Kolmogorov, was defined to be the length of the shortest program with which a general-purpose computer can generate the string. The resulting algorithmic theory of information turned out to have far-reaching implications not only for the theory of computation but also with regard to provability in logical systems, foundations of probability, and inductive inference. However, the theory gives no guidance as to the practical construction of programs, let alone the shortest one, which in fact cannot be found by algorithmic means. Accordingly, the theory has had little or no direct impact on statistical problems.

Inspired by the algorithmic theory of information or complexity, while being aware of an earlier special case discussed by Wallace and Boulton, I developed the *MDL* (Minimum Description Length) principle, originally meant as a criterion for the estimation of the number of parameters. The basic idea was that one cannot encode data with a short code length without taking advantage of the restrictions in the data. The restrictions are described by statistical models, which themselves must be encoded. Its early rough implementation produced a criterion which is equivalent to the Bayes Information Criterion (*BIC*). Unfortunately, this very special instance has given the false impression that this is all the *MDL* principle amounts to. In fact, the principle is very general, somewhat in the spirit of Ockham's razor, although the latter selects the simplest explanation among equivalent ones, while the *MDL* principle applies to all "explanations" of data whether equivalent or not.

This book, which grew out of lecture notes, is aimed at describing the latest developments on the *MDL* principle, which once more were inspired by the unpublished work of the great Kolmogorov in the algorithmic theory of complexity, called his *structure* function. This work not only incorporates the *MDL* principle in an abstract manner but makes clear the fundamental ideas in modeling. Since, again, the notions require the Kolmogorov complexity, they all are non-computable. By replacing the class of models as the set of programs in a universal language by families of probability models, we avoid the non-computability problem and obtain a fairly comprehensive theory of modeling which, unlike the traditional attempts at theory, has no logical fallacies.

A full treatment requires a lot of technical difficulties, and to try to emphasize the ideas we give the proofs of theorems (with one exception) when they are short, while referring to the literature for the details of more complicated and lengthy proofs. Also we do not hesitate to give important statements and relationships without proofs or with sketches, with the understanding that such proofs under reasonable conditions can be given. This is done in order to keep the presentation simple. Still, the level is something like a graduate level, and knowledge of basic probability theory and statistics is needed. On the other hand, the required facts of basic coding theory will be given.

1

Introduction

Statistical modeling or model building is an activity aimed at learning rules and restrictions in a set of observed data, proverbially called "laws" or "the go of it," as stated by Maxwell. The traditional approach, presumably influenced by physics, is to imagine or assume that the data have been generated as a sample from a population, originally of a parametrically defined probability distribution and, later, more generally, a so-called nonparametric distribution. Then the so-imagined unknown distribution is estimated from the data by use of various principles, such as the least-squares and maximum-likelihood principles, or by minimization of some mean loss function, the mean taken with respect to the "true" data generating distribution imagined.

Although such a process can work well if the situation is similar to that in physics – namely, that there is a "law" which is guessed correctly and which is capable of describing the data sufficiently well for one to be able to account for the unavoidable deviations to small random instrument noise. However, in statistical applications the "law" assumed is statistical, and there is no reliable, let alone absolute, way to decide whether a given data set is a sample typically obtained from any suggested distribution. It is true that in some very special cases, such as coin flipping, the physical data-generating mechanism suggests a good probability model, and we do have a reasonable assurance of a good match. In general, however, we do not have enough knowledge of the machinery that generates the data to convert it into a probability distribution, whose samples would be statistically similar to the observed data, and we end up with the impossible task of trying to estimate something that in reality does not even exist. And because we certainly can imagine more than one target distribution to be estimated, an assessment of the estimation error is meaningless. It is of course another matter to analyze estimators and their errors in samples of various distributions, but that belongs to probability theory, and may provide useful information for statistical problems. However, such an activity cannot provide a foundation for statistics.

The assumption of the "true" data-generating distribution is deep-rooted. Even such a well-meaning statement as "all models are wrong but some are useful" is meaningless unless some model is "true". We elaborate this important issue a bit more. Not even in the simple case of coin flipping is there inherent probability associated with the coin; the probabilities are just mathematical models not to be confused with the reality. In a similar vein, the well-known question posed about the length of the shore line of Britain is meaningless without our specifying the way the length is to be measured; i.e., without giving a model of the shore line.

The confusion of the models and reality can have dangerous implications. When the problem of statistics is viewed as an estimation of a probability distribution from data, there is no rational way to compare the different models that could be assumed. Since in general a more complex model "fits" the data better than a simpler one, where the degree of the "fit" is measured in any reasonable way, it is impossible within such a theory to rule out the inevitable conclusion that the best model is the most complex one assumed. The usual way to avoid this disastrous conclusion is to resort to judgment or ad hoc procedures, where the elusive and undefined complexity is penalized in some manner. While there exists a theory for estimating the real-valued parameters, based on Fisher's work, no theory exists for estimating such important parameters as the number of the real-valued parameters or, more generally, the model structure. Good practitioners, of course, know intuitively that one can never find the "true" data-generating distribution, so that it must be regarded only as an unreachable model. But this leads us to the awkward position that we must estimate this model, and since the estimates define the actual model we must live with we end up modeling a model! A particularly difficult modeling problem is the so-called de-noising problem, which we discuss later. The usual assumption that the data consist of a "true" signal, which is corrupted by "noise", does not define noise in any concrete data-dependent form. As a consequence the real definition of noise in such approaches amounts to "that part in the data that is removed"!

In this book we describe a different approach to modeling problems, whose origin is the primitive form of the *MDL* (Minimum Description Length) principle, [50, 53], the basic idea of which in a particular case was proposed earlier in [81]. After a circuitous evolution [58], it now gets expanded in the spirit of Kolmogorov's unpublished work on the *structure function* [79], together with the ideas of "distinguishable" distributions by Balasubramanian [3]. The fundamental premise is that there is no assumption about the data-generating machinery. Instead, we intend to learn the regular features in the data by fitting to them probability models in a selected class. The model class serves simply as a language in which the properties we wish to learn can be described. The objective is to extract a model which incorporates all the properties that can be learned with the model class considered. This involves formalization of the intuitive ideas of "learnable information," "complexity," and "noise," all of which can be measured with the common unit "bit". The work is completed when we can make the statement that "the given data have x bits of learnable information, y bits of complexity, leaving z bits of noise which has no structure to be learned with the model class considered". We illustrate the issues with a simple example. Consider the alternating binary string 010101... as the data. In light of the Bernoulli class with the probability of each symbol one half this string is all noise, and the extracted useful information is zero. On the other hand, a first-order Markov model with the conditional probability of symbol 1 unity at state 0, and of symbol 0 unity at state 1, will render the entire string as useful information, leaving nothing unexplained as noise. We see that two different model classes may extract entirely different properties from the data and impose very different constraints, the learning of which is the very purpose of all modeling.

It is evident that no assumption of a "true" data-generating distribution is needed which the models try to estimate. Hence, the models are neither "true" nor "false" nor "right" nor "wrong". They just have different capabilities, which we can measure; and we can compare and assess different suggestions for the model classes. Now, one might say that we are leaving out of the theory the crucially important task of selecting the model class, but this is exactly what has to be done, for there is a theorem due to Kolmogorov that the selection of the best model and the best model class is noncomputable. Hence finding these must be left for human intuition and intelligence – forever.

There are two ways to formalize the modeling approach outlined. The first was done by Kolmogorov in his structure function in the algorithmic theory of complexity. This theory was introduced by Solomonoff [71] to provide a logical basis for inductive inference, and independently by Kolmogorov [34, 35] and Chaitin [9] as the algorithmic theory of information (the name was changed later to the "algorithmic theory of complexity." For an extensive treatment of the theory we refer to [38]. The theory provides a different foundation for statistical inquiry, one which is free from the problems discussed above. What is most important is that data need not be regarded as a sample from any distribution, and the idea of a model is simply a computer program that describes or encodes the data. This may seem strange, but it can be made to be equivalent with a probability distribution, constructed from the length of the program as the number of binary digits required to describe the program. The length of the shortest program serves as a measure of the complexity of the data string, and there is even a construct, known as Kolmogorov's minimal sufficient statistic, associated with his structure function, which provides a natural measure for the complexity of the best model of the data. All this would leave nothing wanting from statistical inquiry except for a fatal problem: The shortest program, which can be argued to provide the penultimate ideal model of the data, cannot be found by algorithmic means, but for the same reason, the basic ideas can be explained quite easily without technical difficulties. We give later a brief account of the elegant algorithmic theory of complexity as needed here.

The second formalization for a class $\mathcal{M}_\gamma$ or $\bigcup_\gamma \mathcal{M}_\gamma$ of parametric models of the kind $f(\cdot; \theta, \gamma)$, which does not involve non-computability problems, requires basic facts of Shannon's coding theory as well as its suitable generalization. Here, the dot represents data of any kind, say, $y^n = y_1, \ldots, y_n$ or $(y^n, x^n) = (y_1, x_1), \ldots, (y_n, x_n)$, where x^n may affect the values y^n, and $\theta = \theta_1, \ldots, \theta_k$ denote real valued parameters, while γ, ranging over some set, specifies the structure in which the parameters lie. In a simple case γ is just the number k of parameters. We need the models to be finitely describable so that they can be fitted to data. Hence, they would not have to be parametric, for they could be defined by an algorithm. However, the algorithms permitted would have to be restricted to avoid the computability problem, and the simplest restriction is to take the models as parametric. This use of the word *parametric* is wider than what is normally meant in traditional statistics, for they include such *nonparametric* models as histograms.

The *MDL* principle calls for calculation of the shortest code length for the data and the parameters, which when minimized becomes a function of the data, and hence ultimately we seek to minimize the code length for the data, relative to a class of models, such as $\mathcal{M}_\gamma$ or $\bigcup_\gamma \mathcal{M}_\gamma$. This makes intuitive sense since we cannot encode data without taking advantage of all the regular features restricting them that the models permit. The basic coding theory, which will be discussed in the first part of this book, implies that a code length $L(a)$ for any finitely describable object a defines a probability $P(a)$; and, conversely, we may take $L(a) = \log 1/P(a)$. This means that to encode the triplet x^n, θ, γ, we need a prior for the parameters. The real-valued parameters must be quantized so that they can be encoded with a finite code length. This is an important and even crucial difference between the *MDL* theory and the Bayesian approaches, in that in the former only prior knowledge that can be described, i.e., encoded, is permitted, not in any vague sense that there is a certain prior distribution on the parameters, but in the concrete sense that the parameter value needed can be encoded in such a manner that it can be decoded. A code length is equivalent with a probability, and there is no need for the awkward Bayesian interpretation that a probability is a degree of belief. For instance, if a parameter value has the prior weight 0.4, the interpretation is simply that it takes a little more than $\log 1/0.4$ bits to encode this parameter value in a decodable manner.

The decodability requirement has important consequences. First, the maximization of the joint probability of the data and the prior maximizes the Bayesian posterior for the chosen prior. In the Bayesian framework, which lacks a clearly stated principle other than the unrestricted use of probabilities, one cannot maximize the posterior over the priors, while the *MDL* principle has no such restrictions. Further, unlike in the Bayesian approaches, where the entire distribution of the prior is generally required before the actual probability of the object needed can be expressed, in the *MDL* theory only the code length of the actual object is required, which may be much simpler to calculate. Indeed, in very complex situations the description of the entire distribution would be difficult, requiring a very long code length to encode it. An example is the problem of finding "scenes" in a gray level image. A scene may be defined as a region inside a closed curve where the gray level is nearly uniform. Rather than attempting to describe a prior for the huge set of all closed curves on the image, which is a hopeless task, one can encode each tentatively selected curve by various contour chasing techniques, and get the code length needed. This illustrates the advantage of having the logically equivalent alternative for a probability, the code length.

One must bear in mind, however, that there is a danger of infinite regression in the *MDL* theory, in particular, when the prior distribution selected from a collection is encoded, because it may require another hyper prior, etc. Each hyper prior adds something to the code length, at least one bit, and it seems that the data cannot be encoded with a finite code length at all. To make the process meaningful, we simply put a constraint to the selection of the hyper priors such that their code lengths get shorter and shorter and that the highest level hyper prior included is so defined that its code length can be ignored. For instance, there is a universal code for the integers, as described below, which gives the code length about $\log n + 2\log\log n$ to the integer n. This is taken as general knowledge so that the associated prior itself need not be described.

To illustrate the *MDL* principle in a preliminary way consider the problem of maximization of the probability of the data with respect to the structure index. For each fixed structure it is accomplished by a *universal* distribution $\hat{f}(x^n; \mathcal{M}_\gamma)$, which has no other parameters than the structure index. This has the property that it assigns almost as large a probability (or density) to the data as any model $f(x^n; \theta, \gamma)$ in the family. There are a variety of such universal models, some of which are discussed below. With a universal model and a code length for the structure index $L(\gamma)$ the optimum index $\hat{\gamma}(x^n)$ is found by the minimization

$$\min_\gamma \log 1/\hat{f}(x^n; \mathcal{M}_\gamma) + L(\gamma) \ .$$

It may be somewhat surprising that the first term, which often is dominant, already has a penalty for the number of parameters, and the second term may sometimes be dropped. This behavior has its explanation in the fact that the universal distribution may be interpreted actually to define a joint probability of the data and the maximizing data-dependent parameters implicitly and quantized optimally. Especially in de-noising problems, the second term cannot be dropped.

After the optimal structure is found, the problem of finding the best-fitting model, specified by the optimally quantized parameter θ, remains, which requires a deeper theory and will occupy the main part of this book. A particular quantization gives optimally "distinguishable" models, which provide a new and straightforward way to solve the important problems of hypothesis testing and confidence intervals, whose traditional solutions are awkward and fragmentary. In the original crude version of the *MDL* principle, the quantization of the real-valued parameters gave a criterion for the selection of the number of parameters as well, which is equivalent to the criterion called *BIC*. Such a criterion does not take into account important information about the data and is not particularly effective except asymptotically. Since there is no overall guiding objective in Bayesianism the criterion *BIC* has led to a dead end, while the same criterion based on the *MDL* principle has been improved; see the applications of the *NML* model in Chapter 10.

To summarize the main features of the *MDL* theory, the first is philosophical in that no assumption of a "true" data-generating distribution is needed. This changes the objective and foundation for all model building. Instead of trying to minimize this or that criterion to get a model which is close to the "true", and in fact nonexistent, target distribution, the objective is to extract all the useful and learnable information from the data that can be extracted with a model class suggested. The concept of "learnable information" is formally defined, which in many cases but not always creates a special data-dependent criterion, namely, the amount of information measured in terms of code length, which allows for comparison of any two models. This means that both types of parameters, the real-valued ones and their numbers and even the structure in which the parameters lie, are governed by a common theory. One should bear in mind, however, that the *MDL* theory provides a broad principle rather than an isolated criterion to be applied blindly. It is not always possible to find an easily computed closed form criterion to implement it, and, moreover (and this appears to be confusing), one may find several criteria, one sharper than another. After all, in model building we are up against the fundamental problem in science, which is to learn from nature, and a successful application of the *MDL* principle requires a measure of thinking.

Part I

Information and Coding

A formal notion of *information* was introduced at about the same time (around the late 1940s) independently by both Norbert Wiener and Claude Shannon [68, 85]. Despite the fact that a case could be made for this notion of information to be among the most influential ideas in the recent history of science, Wiener did not seem to regard it as worthwhile to elaborate it much further. He gave a brief sketch of its application to communication problems, but that is about it. By contrast, it was Shannon who did provide such an elaboration, and the result was the beginning of an entirely new discipline, information theory, which has had a growing impact in almost all branches of science. As will be clear later on, the basic information measure in a restricted form was already given by Hartley in 1928 [29], and in essence the same concept as a measure of disorder has been used in statistical mechanics for much longer (by Boltzmann [8]).

Initially, information theory was synonymous with communication theory, but gradually the central concept of entropy with variations such as the Kolmogorov–Sinai entropy for chaotic processes started to find important applications in probability theory and other parts of mathematics. In the mid 1960s an exciting new idea emerged, which added a new branch to information theory with goals and theorems completely different from those of communication problems. In this, the information in a string is formalized as the length of the shortest computer program that generates the string. The new measure is usually called the Kolmogorov complexity, even though it was introduced in a clear manner years earlier by Solomonoff. The same idea was rediscovered by Chaitin [9], who has made other important contributions to the theory, especially of the meta-mathematics type.

2

Shannon–Wiener Information

2.1 Coding of Random Variables

In coding we want to transmit or store sequences of elements of a finite set $A = \{a_1, \ldots, a_m\}$ in terms of binary symbols 0 and 1. The set A is called an *alphabet* and its elements are called *symbols*, which can be of any kind, often numerals. The sequences of the symbols are called *messages*, or often just data when the symbols are numerals. We begin by defining the *code* as a function $C : A \to B^*$, taking each symbol in the alphabet into a finite binary string, called a *codeword*. These define a binary tree (see Figure 2.1), in which the codewords appear as paths starting at the root node. We are interested in codes that are one-to-one maps so that they have an inverse. The code may be extended to sequences $x = x_1, \ldots, x_n$, also written as C,

$$C : A^* \to B^* \ ,$$

by the operation of *concatenation*: $C(xx_{n+1}) = C(x)C(x_{n+1})$, where xa denotes the string obtained when symbol a is appended to the end of the string x.

Code trees and the concatenation operation belong to the traditional coding theory and are not the only ways to encode sequences, but because the fundamental coding ideas based on them are simple and intuitive, we describe them in this subsection. In reality, all concatenation codes are special cases of *arithmetic codes*, to be described later, which to a large extent are used today, and in fact the traditional quite elegant constructs are no longer needed and are outdated.

We want the extended code to be not only invertible but also such that the codewords $C(i)$ of the symbols $i \in A$ (we often write the symbols as i rather than x_i or a_i) can be separated and recognized in the code string $C(x)$ without a comma. This implies an important restriction, the so-called prefix property, which states that *no codeword is a prefix of another*. Clearly when climbing the tree, always starting at the root, a codeword is found when the

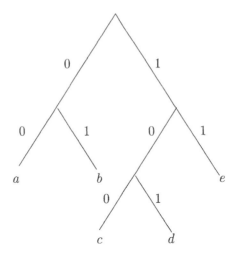

Fig. 2.1. A complete prefix code; alphabet $\{a, b, c, d, e\}$.

leaf is reached and no comma is needed. This requirement, making a code a
prefix code (Figure 2.1) implies the important Kraft inequality,

$$\sum_{i \in A} 2^{-n_i} \leq 1 \,, \qquad (2.1)$$

where $n_i = |C(i)|$ denotes the length of the codeword $C(i)$. If, moreover, each
leaf defines a codeword, the code is called *complete*.

To prove the inequality, consider the leaves of any complete binary tree –
i.e., a tree where each node has either two sons or none. Any such tree can
be obtained by starting with a two-leaf tree and splitting the leaf nodes suc-
cessively until the tree is obtained. For the two-leaf tree the Kraft inequality
holds with equality: $2^{-1} + 2^{-1} = 1$. Splitting a node w of length $|w|$, the
equality $2^{-|w0|} + 2^{-|w1|} = 2^{-|w|}$ holds for the son nodes, just as the proba-
bilities of a Bernoulli process, which by an easy induction implies that the
Kraft inequality holds with equality for any complete binary tree. The codewords of a prefix code are leaves of a tree, which can be completed. Hence,
the claim holds. It is easily extended even to countable alphabets, where the
sum becomes the limit of strictly increasing finite sums bounded from above
by unity.

The codeword lengths of a prefix code, then, define a probability distribution by $P(i) = 2^{-n_i}/K$, where K denotes the sum of the left hand side
of the Kraft inequality. Even the converse is true in essence. Indeed, assume
conversely that a set of integers $n_1, \ldots, n_m$ satisfy the Kraft inequality. We
shall describe a prefix code with the lengths of the codewords given by these
integers. Let $\hat{n}$ be the maximum of these integers. Write each term as

$$2^{-n_i} = 2^{r_i} \times 2^{-\hat{n}} \,,$$

where $r_i = \hat{n} - n_i$, and

$$\sum_i 2^{-\hat{n}} 2^{r_i} \leq 1 \,,$$

by the assumption. This is a sum of the leaf probabilities $2^{-\hat{n}}$ over a subset of the leaves $0...0, 0...01, ...$ of a balanced tree of depth $\hat{n}$. Hence, we get the required prefix code by partitioning this subset into m segments $2^{r_1}, 2^{r_2}, ...$ of adjacent leaves in each such that each segment has a common mother node of length n_i, which we take as the codeword.

A practical coding problem of some importance results when we ask how to design a prefix code when the symbols are taken independently with probabilities $P(a)$, and we want the code to have as short a mean length,

$$L^*(P) = \min_{\{n(a_i)\}} \sum_i P(a_i) n(a_i) \,,$$

as possible, where the lengths $n(a_i)$ are required to satisfy the Kraft inequality. This may also be regarded as a measure of the *mean complexity* of the alphabet A or its symbols taken by sampling the distribution P. Indeed, intuitively, we consider an object "complex" if it takes a large number of binary digits to describe (= encode).

Optimal lengths and a corresponding prefix code can be found by Huffman's algorithm, which we describe because of its elegance and for historical reasons. First by relabeling, if necessary, let the probabilities be sorted by nonincreasing size $p_1 \geq p_2 \geq \ldots \geq p_n$. Then, clearly, in the optimal prefix code the codeword lengths must be nondecreasing $\ell_1 \leq \ell_2 \leq \ldots \leq \ell_n$. Indeed, if for some pair $\ell_i > \ell_{i+1}$, then $p_i \ell_{i+1} + p_{i+1} \ell_i < p_i \ell_i + p_{i+1} \ell_{i+1}$, which contradicts the optimality.

Further, for optimality the prefix code must be complete; i.e., every leaf is a codeword, because otherwise at least one codeword at its leaf could be shortened by moving it to the father node. Consider a deepest leaf from the root of length ℓ_n. Its sibling has the same length and has to be assigned a codeword. Hence $\ell_n = \ell_{n-1}$, and the probability of their father node is marked with $p_n + p_{n-1}$. We now need to construct an optimal prefix tree for an alphabet of size $n-1$, with the original symbols and their probabilities except that the last two symbols are lumped together with the probability $p_n + p_{n-1}$. The new symbol has a codeword of length $\ell_n - 1$ in the reduced alphabet. Sort the sequence of the $n-1$ probabilities to a nonincreasing sequence, and assign two leaves as siblings to the two longest codewords. Mark their common father node with the sum of their probabilities. Continue in this fashion by reducing the alphabet, and when the sum of the leaf probabilities of an already constructed father node becomes the smallest or the next smallest probability a new father node gets created in a reduced alphabet, and hence an internal node in the final tree. We are done when the alphabet is reduced to just two symbols and their father node is the root with probability unity.

Example

The sequence of probabilities is 0.3, 0.3, 0.2, 0.1, 0.1. The first reduction of the alphabet gives the sorted probabilities 0.3, 0.3, 0.2, 0.2; the second, 0.4, 0.3, 0.3; and the third, 0.6, 0.4, which gives a tree defined by the codewords 11,10,00,010,011 with the originally given probabilities. The tree constructed is not unique nor even the sequence of the codeword lengths. Only the mean length is minimal.

A far more important result than the optimal tree construction is the following fundamental theorem in information theory:

Theorem 1 *For any code for which the Kraft inequality holds, the mean code length $L(P)$ satisfies*

$$(i)\ L(P) \geq -\sum_i P(a_i) \log_2 P(a_i) \equiv H(P)$$

$$(ii)\ L(P) = H(P) \Leftrightarrow n(a_i) \equiv -\log_2 P(a_i)\,,$$

where $0 \log_2 0 = 0$.

Proof. We give a very simple proof of this fundamental theorem, called often Shannon's noiseless coding theorem, but in reality it is due to McMillan and Doob (as told me by Sergio Verdu). Letting log stand for the binary logarithm and ln for the natural one, we have

$$\sum P(a_i) \log \frac{2^{-n(a_i)}}{P(a_i)} = (\log e) \sum P(a_i) \ln \frac{2^{-n(a_i)}}{P(a_i)}$$

$$\leq (\log e) \sum P(a_i)[\frac{2^{-n(a_i)}}{P(a_i)} - 1]$$

$$= (\log e)[\sum 2^{-n(a_i)} - 1] \leq 0\,.$$

The first inequality follows from $\ln x \leq x - 1$.

The lower bound $H(P)$ of the mean code lengths in the theorem is the famous *entropy*, which may be taken as a measure of the *ideal* mean complexity of A in light of P. Indeed, $H(P) \leq L^*(P) \leq H(P)+1$, and for large alphabets the entropy is close to $L^*(P)$. The *ideal* code length $-\log P(a_i)$, the *Shannon–Wiener information*, may be taken as the *complexity* of the element a_i, relative to the given distribution P. The name "ideal" simply suggests that the integer length requirement for a real code length is often quite irrelevant, and things get simpler by omitting it. The role of the distribution P is that of a *model* for the elements in the set A in that it describes a property of them, which in turn, restricts the elements in a collective sense rather than individually. Entropy, then, measures the strength of such a restriction in an inverse manner. For instance, binary strings generated by sampling a Bernoulli distribution (or *source*) such that the probability of symbol 0 is much greater than that of 1 have a low entropy meaning a strong restriction and allowing for a short mean code length. Conversely, coin flipping strings have little that constrains them, and they have the maximum entropy 1 per symbol.

The meaning of the abstract idea of entropy is often explained as measuring the amount of uncertainty, which, however, is another abstract word. Actually the uncertainty involved is exactly in the same sense as elements in a large set, which have greater uncertainty than elements in a small set. And this of course is reflected in the code length. Suppose we consider encoding strings of length n, where symbol a_i occurs n_i times. No matter how we assign the codeword to the symbol a_i the code length of most such strings cannot be less than $\log(n!/\prod_i n_i!)$, because this is the logarithm of the total number of strings with the given counts. If we set the code words of all the strings as nodes in a balanced binary tree, which of course would be more efficient than coding with any prefix code, only one-half of the strings could be encoded with a shorter code length than the maximum minus one, one-quarter than the maximum minus two, and so on. Also the mean code length calculated with the probabilities n_i/n would be only slightly less than the maximum. Finally, when we apply Stirling's approximation formula

$$k! \cong \left(\frac{k}{e}\right)^k \sqrt{2\pi k}$$

to the factorials we see that

$$\frac{1}{n}\log\frac{n!}{\prod_i n_i!} \to H(\{n_i/n\})$$

as n grows.

There is also already a hint in the theorem of the intimate connection between code length minimization and model selection, which will be the main theme in this book. To see this, consider the class of independent identically distributed (iid) processes over the alphabet A, defined by the parameters $P(a_i)$ for $a_i \in A$. Suppose we have a data sequence from A^n of length n generated by some unknown process in the family, and we consider encoding the data sequence with a prefix code. If we assign a binary string of length ℓ_i to the symbol a_i, the code length for the string would be $\sum_i n_i \ell_i$, where n_i denotes the number of times symbol a_i occurs in the string. The code length would be minimized for $\ell_i = \log(n/n_i)$, which can be seen by the theorem, except that ℓ_i will have to be an integer. We can take $\ell_i = \lceil \log(n/n_i) \rceil$ to satisfy the Kraft inequality, and we would get a near optimal code length, at least for large alphabets, because the extra length amounts at most to one bit per symbol. However, if we do not worry about the integer length requirement and consider $\log(n/n_i)$ as an ideal code length, we see that the optimal ideal code length gives the maximum likelihood estimate of the unknown parameters $P(a_i)$ of the process that generated the data.

2.1.1 Universal prior for integers and their coding

A good example of the equivalence of prefix code lengths and distributions is the problem of designing a prefix code for the integers, which in traditional terminology would be called a "prior". Unlike in Bayesian thinking, we cannot put just any prior probability to an integer like $n = 210$ or 11010010 in binary, because it will have to be decodable even when this binary string is followed by other binary symbols, as the case will be in applications. The basic idea is to attach a binary number to tell the decoder that the following eight symbols represent the integer. But how does the decoder know that the indicated number 8, or 1000 in binary, approximately $\log 210$ symbols, in the string represents the length of the binary integer n? We need to attach a string of length 3, approximately $\log \log 210$ bits, in front to tell the decoder that the following 3, approximately $\log \log \log 210$ symbols, give the length of the number of bits in the binary number 8, or the number 4, and so on. To describe the length of the given integer requires then approximately $\log 210 + \log \log 210 + \log \log \log 210$ bits. An implementation of a real code along this idea was done by Elias [18] in Table 2.1, which gives the length of the integers to be encoded:

Table 2.1. Coding table for the length of integers.

Length ℓ	Code
2	1 0 10 1
3	1 0 11 1
4	1 0 11 0 100 1
5	1 0 11 0 101 1
6	1 0 11 0 110 1
7	1 0 11 0 111 1
8	1 0 11 0 100 0 1000 1
.	. . .

Only integers greater than 1 will have length information. The symbol 0 after each length number indicates that the subsequent symbols describe a length, while 1 indicates that the subsequent bits describe the actual number. The coding of the length number begins with 1. The first 1 followed by 1 encodes the integer 1, and the first 1 followed by 0 tells that this is not the final number, but the two bits tell the length of another length number, and so on. The number 210 or 11010010 is then encoded as 101101000100011010010. Clearly, this code has redundancies and it is needlessly long for small integers. However, we are really interested in the distribution function for the natural numbers, whose negative logarithm gives the number of bits required to encode an integer, rather than the actual code.

Let $\log^k(n)$ denote the k-fold composition of the binary logarithm function, and let exp^k denote its inverse. For example, $exp^0(x) = x$, $exp^1(x) = 2^x$, $exp^2(2) = exp(exp(2)) = 2^4 = 16$, and $exp^3(2) = 2^{16}$. We also write $e(k) = exp^k(2)$. Further, let $log^*(n) = \log n + \log \log n + \ldots$, where the last term is the last positive k-fold composition. We wish to calculate the sum $c = \sum_{n>0} 2^{-\log^* n}$, as in [53], more accurately than the upper bound given in [40].

The derivative of the k-fold logarithm function is given by

$$D \log^k(x) = 1/[a^k x \log x \times \ldots \times \log^{k-1}(x)] = a^{-k} 2^{-\log^* x}$$

for $e(k-2) \leq x \leq e(k-1)$ and $e(i) = 0$ for $i < 0$, where $a = \ln 2 = 0.69....$
We can evaluate the integral

$$\int_{e(k-1)}^{e(k)} 2^{-\log^* x} dx = a^{k+1} ,$$

which further gives

$$\int_{e(k)}^{\infty} 2^{-\log^* x} dx = a^{k+2}/(1-a) = S(k) .$$

Next

$$2^{-\log^* n} < \int_{n-1}^{n} 2^{-\log^* x} dx < 2^{-\log^*(n-1)} , \qquad (2.2)$$

which implies

$$\sum_{1}^{n-1} 2^{-\log^* i} + S(n) < c < \sum_{1}^{n-1} 2^{-\log^* i} + S(n-1) . \qquad (2.3)$$

The difference between the two bounds is by (2.2) less than $2^{-\log^*(n-1)}$. By putting $n-1 = 16 = e(2)$ and calculating the upper bound from (2.3) we get $c = 2.86$ with error less than 2^{-7}. By putting $n-1 = 2^{16} = e(3)$ and calculating the upper bound from (2.3) we get $c = 2.865064$.

Put $w^*(n) = c^{-1} 2^{-\log^* n}$, which defines a *universal prior* for the integers. It gives the (ideal) code length

$$L^*(n) = \log 1/w^*(n) = \log^* n + \log c . \qquad (2.4)$$

Let $L(n)$ be any length function satisfying the Kraft inequality. As in [12], consider the set of integers $A = \{n : L(n) \leq L^*(n) - k\}$, which is the same set as $\{n : w^*(n) \leq 2^{-L(n)-k}\}$, and let $I_A(n)$ be its indicator function, which is 1 if $n \in A$ and zero otherwise. Then

$$\sum_n w^*(n) I_A(n) \leq \sum_n I_A(n) 2^{-L(n)-k}$$

$$\leq \sum_n 2^{-L(n)-k} \leq 2^{-k} .$$

We see that the probability of the set of integers for which any prefix code length function can beat $L^*(n)$ by more than k bits decreases exponentially in k. This can be augmented by another result as follows. The uncertainty in the set of integers as measured by the entropy $\sum_n w^*(n) \log 1/w^*(n)$ is infinite. We prove the theorem –

Theorem 2 *Let the probability distribution P satisfy*

$$P(i) \geq P(i+1), \ i > M, \ some \ M \tag{2.5}$$
$$H(P) = \infty . \tag{2.6}$$

Then

$$\lim_N \frac{\sum_{n \leq N} P(n) L^*(n)}{\sum_{n \leq N} P(n) \log 1/P(n)} = 1 .$$

Proof. Write $P_N = \sum_1^N P(i)$ and $L_N^* = \sum_1^N P(i) L^*(i)$. Similarly, let $H_N = \sum_1^N P(i) \log 1/P(i)$. By the noiseless coding theorem,

$$\sum_{i=1}^N P(i) \log \frac{w_N^* P(i)}{P_N w^*(i)} = L_N^* - H_N + P_N \log(w_N^*/P_N) \geq 0 .$$

Because $\epsilon \log \epsilon \to 0$, as $\epsilon \to 0$, the third term goes to zero as $N \to \infty$, and

$$\lim_N L_N^*/H_N \geq 1 . \tag{2.7}$$

We have by Wyner's inequality [86], $j \geq 1/P(j)$,

$$H_N \geq \sum_{j=1+M}^N P(j) \log 1/P(j) \geq \sum_{j=M+1}^N P(j) \log j . \tag{2.8}$$

Further,

$$L_N^* \leq \sum_{j=M+1}^N P(j) \log j + \sum_{j=M+1}^N P(j) r(j) + C ,$$

where $r(j) = L^*(j) - \log j$, and C is the maximum of $L^*(j)$ for $j \leq M$. Let $f(j) = r(j)/\log j$, and let $K(\epsilon)$ be an index such that $K(\epsilon) > M$ and $f(j) \leq \epsilon$ for all $j \geq K(\epsilon)$. Then for $N > K(\epsilon)$,

$$\sum_{j=M+1}^N P(j) r(j) \leq \epsilon \sum_{j=K(\epsilon)+1}^N P(j) \log j + R(\epsilon) ,$$

where $R(\epsilon)$ denotes the maximum value of $r(i)$ for $i \leq K(\epsilon)$. With this inequality and (2.8) we get

$$L_N^*/H_N \leq 1 + \epsilon + [R(\epsilon) + C]/H_N .$$

This holds for all ϵ, and when $N \to \infty$ the theorem follows.

Often there is an upper bound for the integers given, in which case we may get a shorter code length than $L^*(k)$, especially for small integers. Take

$$w(k|n) = \frac{1/k}{\sum_{i=1}^n 1/i} .$$

We can upper-bound the sum by

$$S = \sum_{i=1}^n 1/i < 1 + \int_1^{n-1} 1/x \, dx = 1 + \ln n ,$$

and we get

$$L(k|n) = \log k + \log \ln(en) . \tag{2.9}$$

2.2 Basic Properties of Entropy and Related Quantities

For these traditional results we follow the excellent book by Thomas and Cover [12], where many others are discussed. Notation: when a random variable (r.v.) X has the distribution P, which is symbolized as $X \sim P$, we write frequently $H(P)$ as $H(X)$.

Clearly,

(1) $H(X) \geq 0$,

because it is the sum of non-negative elements. In the important binary case, where $P = \{p, 1 - p\}$, we write $H(P) = h(p)$. It is a symmetric concave function of p, reaching its maximum value 1 at $p = 1/2$, and vanishing at points 0 and 1.

(2) If $(X, Y) \sim p(x, y)$, then the *conditional* entropy is defined as

$$H(Y|X) = \sum_{x,y} p(x, y) \log \frac{1}{p(y|x)}$$

$$= \sum_x p(x) \sum_y p(y|x) \log \frac{1}{p(y|x)}$$

$$= \sum_x p(x) H(Y|X = x) .$$

Notice that $H(X|X) = 0$ and $H(Y|X) = H(Y)$, if X and Y are independent.

We have

(3) $H(X, Y) = H(X) + H(Y|X)$.

Prove the following:

(4) $H(X,Y|Z) = H(X|Z) + H(Y|X,Z)$.

The *relative entropy* between two distributions p and q is defined as

$$D(p\|q) = \sum_x p(x) \log \frac{p(x)}{q(x)} .$$

By the noiseless coding theorem,

(5) $D(p\|q) \geq 0$,

equality iff $p(x) \equiv q(x)$.

The relative entropy is also called the Kullback–Leibler distance. Note that $D(p\|q) \neq D(q\|p)$ in general.

The entropy has a distinctive grouping property: If we lump together a part of the symbols, say, in a subset $S = \{a_1, \ldots, a_k\}$ of the alphabet $A = \{a_1, \ldots, a_n\}$, which has the distribution $p = \{p(a_i)\}$, then

$$H(p) = h(P_S) + P_S H(\{p(a_1)/P_S, \ldots, p(a_k)/P_S\})$$
$$+(1 - P_S)H(\{p(a_{k+1})/(1 - P_S), \ldots, p(a_n)/(1 - P_S)\}) ,$$

where $P_S = \sum_{i=1}^k p(a_i)$. Hence, the entropy remains the same when we add to the entropy of the binary event S the entropies of the conditional events defined by the outcomes within S and A/S. The interested reader may want to verify this.

The important *mutual information* is defined as

$$I(X;Y) = \sum_x \sum_y p(x,y) \log \frac{p(x,y)}{p(x)q(y)} = D(p(X,Y)\|p(X)q(Y)),$$

where the two marginals are denoted by $p(x)$ and $q(y)$, respectively.

Because $p(x,y) = p(y,x)$ the mutual information is symmetric: $I(X;Y) = I(Y;X)$.

Further,

$$I(X;Y) = H(X) - H(X|Y) = H(Y) - H(Y|X) .$$

To see this, write

$$I(X;Y) = \sum_{x,y} p(x,y) \log \frac{p(x|y)q(y)}{p(x)q(y)}$$

$$= \sum_x [\log \frac{1}{p(x)}] \sum_y p(x,y) - \sum_{x,y} p(x,y) \log \frac{1}{p(x|y)}$$

$$= \sum_x p(x) \log \frac{1}{p(x)} - \sum_x p(x) \sum_y p(x|y) \log \frac{1}{p(x|y)} .$$

It follows from (5) that

(6) $I(X;Y) \geq 0$,

and the equality if X and Y are independent. More generally, the *conditional mutual information* is defined as

$$I(X;Y|Z) = H(X|Z) - H(X|Y,Z) \,,$$

which also satisfies

(7) $I(X;Y|Z) \geq 0$,

the equality holding if X and Y are conditionally independent. Indeed,

$$I(X;Y|Z) = \sum_{x,y,z} p(x,y,z) \log \frac{p(x,y|z)}{p(x|z)p(y|z)} \,,$$

which vanishes exactly when the numerator can be written as the denominator.

It further follows from (5) that

(8) $H(X) \leq \log |A|$,

the equality if X has the uniform distribution $u(x) = 1/|A|$, where A denotes the range of X and $|A|$ its size. Indeed,

$$D(p\|u) = \sum_x p(x) \log \frac{p(x)}{u(x)} = \log |A| - H(X) \geq 0 \,.$$

Equation (6) implies at once

(9) $H(X|Y) \leq H(X)$,

the equality holding if X and Y are independent.

2.3 Channel Capacity

By a "channel" information theorists mean a conditional probability distribution $p(y|x)$ to model a physical channel, where x denotes random symbols, represented by binary strings, entering the channel and y the usually related but in general distorted symbols exiting the channel. By coding we have the chance of selecting the probability distribution by which the input symbols are put into the channel, say, $w(x)$. The problem of utmost importance is to select the distribution $w(x)$ so that the mutual information is maximized:

(10) $\max_w I_w(X;Y) = \max_w \sum_x w(x) D(p(Y|x)\|p(Y))$,

where, we realize, $p(y) = p_w(y) = \sum_x p(x,y) = \sum_z p(y|z)w(z)$ depends on w. This is what makes the maximization problem difficult. We can, however, derive an important necessary condition for the maximizing distribution $w^*(x)$:

(11) $D(p(Y|x)\|p_{w^*}(Y))$

is the same for all x, and hence the *channel capacity* C is given by

(12) $C = I_{w^*}(X;Y) = D(p(Y|X)\|p_{w^*}(Y))$.

We derive (11). To find the maximizing probability function $w = \{w_i\}$ in (10), where we write $w_i = w(i)$, form the Lagrangian

$$K(w) = \sum_x w_x \sum_y p(y|x) \log \frac{p(y|x)}{p_w(y)} - \lambda(\sum_x w_x - 1) .$$

Set the derivative with respect to w_x to zero

$$\frac{\partial K}{\partial w_x} = \sum_y p(y|x) \log \frac{p(y|x)}{p_w(y)}$$

$$- \sum_z w_z \sum_y p(y|z) \frac{p(y|x)}{p_w(y)} - \lambda = 0 .$$

If on the second line we take the sum over z first and then over y we get

$$D(p(Y|x)\|p_w(Y)) = \lambda + 1$$

for all x at the maximizing w.

There is an algorithm due to Arimoto and Blahut to calculate the channel capacity. First, it is not too hard to see that the capacity results from the double maximization

(13) $C = \max_{w(x)} \max_{q(x|y)} \sum_x \sum_y w(x)p(y|x) \log \frac{q(x|y)}{w(x)}$.

In fact, for each $w(x)$ consider $g_q(y|x) = q(x|y)p_w(y)/w(x)$, and the $g_q(y|x)$ that maximizes $\sum_y p(y|x) \log g_q(y|x)$ is $p(y|x)$ by the noiseless coding theorem, from which the claim follows. Then by starting with a guess for the maximizing $w(x)$, say, the uniform, we find the maximizing conditional distribution, which is

$$q(x|y) = \frac{w(x)p(y|x)}{\sum_u w(u)p(y|u)} .$$

For this conditional distribution we find the next $w(x)$ by maximization, which can be done easily with Lagrange multipliers. The result is

$$w(x) = \frac{\prod_y q(x|y)^{p(y|x)}}{\sum_x \prod_y q(x|y)^{p(y|x)}} ,$$

and the cycle can be repeated. That it converges to the channel capacity follows from a general result due to Csiszar and Tusnady.

2.4 Chain Rules

Theorem 3 *Let $X_1, \ldots, X_n \sim p(x_1, \ldots, x_n)$. Then*

$$H(X_1, \ldots, X_n) = \sum_{i=1}^{n} H(X_i | X_{i-1}, \ldots, X_1) \,.$$

Proof. For $n = 2$ the claim follows from (3). For $n = 3$, we get first from (4)

$$H(X_1, X_2, X_3) = H(X_1) + H(X_2, X_3 | X_1) \,,$$

and then the claim follows with an application of (4) to the second term. By a simple induction the claim follows for any n.

Theorem 4

$$I(X_1, \ldots, X_n; Y) = \sum_{i=1}^{n} I(X_i; Y | X_{i-1}, \ldots, X_1) \,.$$

Proof. The claim follows from the previous theorem and the definition of the conditional mutual information. For instance, for $n = 2$

$$I(X_1, X_2; Y) = H(X_1, X_2) - H(X_1, X_2 | Y)$$
$$= H(X_1) + H(X_2 | X_1) - H(X_1 | Y) - H(X_2 | X_1, Y),$$

the claim follows by taking the first and the third terms and the second and the fourth, and adding them together.

Finally, define the conditional relative entropy as

$$D(p(Y|X) \| q(Y|X)) = \sum_x p(x) \sum_y p(y|x) \log \frac{p(y|x)}{q(y|x)} \,.$$

Theorem 5

$$D(p(Y, X) \| q(Y, X)) = D(p(X) \| q(X)) + D(p(Y|X) \| q(Y|X))$$

Proof. Immediate from the definitions:

$$D(p(Y, X) \| q(Y, X)) = \sum_{x,y} p(x, y) \log \frac{p(x)p(y|x)}{q(x)q(y|x)} \,,$$

which gives the claim.

2.5 Jensen's Inequality

An important tools in information theory is Jensen's inequality. A function $f(x)$ is called *convex* in an interval (a, b), if for x and y in the interval

$$f(\lambda x + (1 - \lambda)y) \le \lambda f(x) + (1 - \lambda)f(y)$$

for all $0 \le \lambda \le 1$; i.e., if $f(x)$ does not lie above any cord. If it lies below, except for the two points, the function is *strictly convex*. A function $f(x)$ is concave if $-f(x)$ is convex.

Theorem 6 *If $f(x)$ is convex, then the mean, written as the operation "E", satisfies*

$$Ef(X) \ge f(EX).$$

If f is strictly convex, the equality implies that X is constant.

Proof. (For discrete X.) Let the range of X be just the two points x_1 and x_2 with $p_i = P(x_i)$. Then by convexity

$$p_1 f(x_1) + p_2 f(x_2) \ge f(p_1 x_1 + p_2 x_2).$$

Let the theorem then hold for $k - 1$ points. We have

$$\sum_{i=1}^{k} p_i f(x_i) = p_k f(x_k) + (1 - p_k) \sum_{i=1}^{k-1} \frac{p_i}{1 - p_k} f(x_i)$$

$$\ge p_k f(x_k) + (1 - p_k) f\left(\sum_{i=1}^{k-1} \frac{p_i}{1 - p_k} x_i\right)$$

$$\ge f[p_k x_k + (1 - p_k) \sum_{i=1}^{k} \frac{p_i}{1 - p_k} x_i] = f\left(\sum_{i=1}^{k} p_i x_i\right).$$

We conclude this section by demonstrating that the relative entropy is a convex function of its arguments, the two probability measures p and q. This further implies that the entropy is a concave function of its argument, the single probability measure. Here, the definitions of convexity and concavity are generalized from functions of the reals to functions of any arguments for which linear combinations may be formed.

Theorem 7 $D(p\|q)$ *is convex:*

$$D(\lambda p + (1 - \lambda)p' \| \lambda q + (1 - \lambda)q') \le \lambda D(p\|q) + (1 - \lambda)D(p'\|q'),$$

for $0 \le \lambda \le 1$.

Proof. First, for two sets of numbers $a_i > 0$ and $b_i > 0$,

$$(i) \quad \sum_i a_i \log \frac{a_i}{b_i} \geq (\sum_i a_i) \log \frac{\sum a_i}{\sum_i b_i} .$$

This is seen to be true by an application of the noiseless coding theorem to the distributions $a_i / \sum_j a_j$ and $b_i / \sum_j b_j$. Hence, by putting $a_1 = \lambda p(x)$, $a_2 = (1-\lambda)p'(x)$, $b_1 = \lambda q(x)$, and $b_2 = (1-\lambda)q'(x)$, applying (i), and summing over x we get the claim.

Theorem 8 $H(p)$ *is concave:*

$$H(\lambda p + (1 - \lambda)p') \geq \lambda H(p) + (1 - \lambda)H(p') ,$$

for $0 \leq \lambda \leq 1$.

Proof. For the uniform distribution $u(x)$ we have

$$H(p) = \log |A| - D(p\|u) ,$$

and since $D(p\|u)$ as a function of p is convex by the previous theorem, the claim follows.

2.6 Theory of Types

We give a brief account of the important theory of types. Consider an independent identically distributed (iid) process over the alphabet A.

Definition: The *type* of a sequence $x = x_1, \ldots, x_n$ is the empirical probability measure P_x on A

$$P_x = \{P_x(a) = \frac{n(a|x)}{n} : a \in A\} ,$$

where $n(a|x)$ denotes the number of times symbol a occurs in x.

Definition: $\mathcal{P}_n$ is the *set* of types in A^n.

As an example, for $A = B$, the binary alphabet,

$$\mathcal{P}_n = \{(0, 1), (\frac{1}{n}, \frac{n-1}{n}), \ldots, (\frac{n-1}{n}, \frac{1}{n}), (1, 0)\} .$$

Definition: type *class* of P_x is

$$\mathcal{T}(P_x) = \{y \in A^n : P_y = P_x\} .$$

As an example, let $A = \{1, 2, 3\}$ and $x = 11321$. Then

$$P_x(1) = 3/5, P_x(2) = P_x(3) = 1/5$$
$$\mathcal{T}(P_x) = \{11123, 11132, \ldots, 32111\}$$
$$|\mathcal{T}(P_x)| = \binom{5}{3, 1, 1} = \frac{5!}{3!1!1!} = 20$$

Theorem 9
$$|\mathcal{P}_n| \le (n+1)^{|A|}$$

Proof. Each P_x is a function $P_x : A \to \{0, 1/n, \ldots, n/n\}$, and there are at most $(n+1)^{|A|}$ functions.

Theorem 10 *Let $\{X_i\}$ be an iid process over A with distribution Q. We denote also by $Q(x) = \prod_{i=1}^{n} Q(x_i)$ the probability of a sequence $x = x^n = x_1, \ldots, x_n$. Then*
$$Q(x) = 2^{-n[H(P_x) + D(P_x \| Q)]}.$$

Proof.

$$Q(x) = \prod_{i=1}^{n} Q(x_i) = \prod_{a \in A} Q^{n(a|x)}(a)$$
$$= \prod_{a \in A} Q^{n P_x(a)}(a) = \prod_{a \in A} 2^{n P_x(a) \log Q(a)}$$
$$= 2^{-n \sum_{a \in A} P_x(a)[\log \frac{P_x(a)}{Q(a)} - \log P_x(a)]}.$$

Corollary: If $Q = P_x$, then

$$Q(x) = P_x(x) = 2^{-n H(P_x)}.$$

Example. For a Bernoulli process,

$$-\log P(x | \frac{n_0}{n}) = n H(\frac{n_0}{n}) = n \log n - \sum_{i=0}^{1} n_i \log n_i.$$

Theorem 11 *For any $P_x \in \mathcal{P}_n$*

$$\frac{1}{(n+1)^{|A|}} 2^{n H(P_x)} \le |T(P_x)| \le 2^{n H(P_x)}. \qquad (2.10)$$

Proof. Upper bound:

$$1 \ge \sum_{x \in T(P_x)} 2^{-n H(P_x)} = |T(P_x)| 2^{-n H(P_x)}.$$

Lower bound: Since $A^n = \cup_{P_x \in \mathcal{P}_n} T(P_x)$ we have

$$\max |T(P_x)| \ge \frac{|A^n|}{|\mathcal{P}_n|} \ge \frac{|A^n|}{(n+1)^{|A|}}.$$

Also, $|A^n| \ge 2^{n H(P_x)}$, which follows from (8) above.

Theorem 12 *For any $P \in \mathcal{P}_n$ and any Q*

$$\frac{1}{(n+1)^{|A|}} 2^{-nD(P\|Q)} \le Q(x) \le 2^{-nD(P\|Q)}.$$

Proof.

$$Q(\mathcal{T}(P)) = |\mathcal{T}(P)| 2^{-n[H(P)+D(P\|Q)]},$$

which with (2.10) implies the claim.

Corollary: If $Q = P$, then

$$P(\mathcal{T}(P)) = |\mathcal{T}(P)| 2^{-nH(P)} .$$

2.7 Equipartition Property

The information theoretic inequalities for the entropy and the related quantities derived above, which all are properties in the mean, get strengthened when the range of the random variables involved is large, because then the results hold to a close approximation essentially for all values and not just in the mean. After all, that is what we intuitively want. For instance, we may then talk about the complexity of individual strings in the sense that nearly all strings generated by a certain probabilistic source have such a complexity, and so on. Although the alphabet or the range of a random variable often is not large, we get random variables with a large range when instead of symbols we consider sequences of them.

Define the set of typical sequences of an iid process over A with distribution Q as follows:

$$\mathcal{T}_Q^\epsilon = \{x^n = x : D(P_x\|Q) \le \epsilon\},$$

where P_x is the type of x. Then

$$
\begin{aligned}
1 - Q(\mathcal{T}_Q^\epsilon) &= \sum_{P_x:D(P_x\|Q)>\epsilon} Q(\mathcal{T}(P)) \\
&\le \sum_{P_x:D(P_x\|Q)>\epsilon} 2^{-nD(P_x\|Q)} \\
&\le \sum_{P_x:D(P_x\|Q)>\epsilon} 2^{-n\epsilon} \\
&\le (n+1)^{|A|} 2^{-n\epsilon} = 2^{-(\epsilon-|A|\log(n+1))},
\end{aligned}
$$

which goes to zero as $n \to \infty$. Hence, as n grows, the probability of the set of typical sequences goes to one at the near exponential rate, no matter what ϵ is. Moreover, by Theorem 9 all typical sequences have just about equal probability

$$2^{-n(H(Q)+\epsilon)} \leq Q(x_1, \ldots, x_n) \leq 2^{-n(H(Q)-\epsilon)}$$

given by the entropy. This is called the asymptotic equipartition property.

Since the tail probabilities of the complement events of T_Q^ϵ are summable over n, the probability of the limsup of these events is zero by the Borel–Cantelli lemma. Therefore the probability of the set of infinite strings, along which $D(P_{x^n} \| Q(x^n)) \leq \epsilon$ for all but finitely many times, is unity for all positive ϵ. This means that when n reaches some finite number, the fluctuation of the type about the data-generating probability Q does not exceed a fixed amount depending on ϵ, and it goes to zero if we let ϵ shrink to zero. In other words, $n_i(x^n)/n \to Q(i)$, almost surely, where $n_i(x^n)$ denotes the number of the occurrences of symbol i in x^n.

Historical Notes

This chapter provides the standard facts about concatenation codes; see, for instance, [26]. There are intricate classes of codes other than prefix codes, such as "instantaneous" and uniquely decodable but "not instantaneous" codes, but since all concatenation codes have been to a large extent outdated by arithmetic codes, discussed below, we do not discuss them. The prior for integers is a modification of the papers [18] and [53].

We have presented just a sample of the most basic properties of entropy and the related quantities, which we drew from [12], where a wealth of beautiful facts are given. The book also includes deep applications of information theoretic results to probability theory and is warmly recommended to any interested reader.

3

Coding of Random Processes

Up to this point we have considered encoding data modeled as outcomes of random variables, which are defined by an alphabet and a single probability distribution. When such a distribution is sampled repeatedly, we actually generate data by an iid (independent identically distributed) random process. A much more important way to model data sequences is to consider them as samples from a general random process, which we discuss next.

3.1 Random Processes

Definition. A countable sequence of random variables $X_1, X_2, \ldots$ is said to form a *random (or stochastic) process*, *r.p.*, if a (measurable) function $P : A^* \to [0, 1]$, where A^* is the set of all finite strings, exists satisfying the two axioms

1. $P(\lambda) = 1$, λ is the empty string,
2. $\sum_{x_{n+1}} P(x_1, \ldots, x_{n+1}) = P(x_1, \ldots, x_n)$, all n.

For each n the function P generates a joint probability measure $P_n(x_1, \ldots, x_n)$, because the $X_i's$ are random variables. The second axiom requires that the marginal probability function obtained from P_{n+1} on the collection $x_1, \ldots, x_n$ agrees with the joint probability function P_n on the same collection; i.e.,

$$\sum_{x_{n+1}} P_{n+1}(x_1, \ldots, x_{n+1}) = P_n(x_1, \ldots, x_n), \tag{3.1}$$

which of course is not automatically satisfied. When it is true, we can safely omit the subindex n, as we did in the two axioms above, which notation actually conforms with the Kolmogorov extension theorem. This states that any set of probability measures P_n satisfying Equation (3.1) admit a unique extension P, defined for infinite strings. The same axiom permits us to define the conditional probabilities

$$P(x_{n+1}|x^n) = \frac{P(x^{n+1})}{P(x^n)} ,$$ (3.2)

which by linking the past with the future is clearly needed to make meaningful predictions.

The idea of stationarity is captured by the axiom

$$\sum_{x_1} P(x_1, \ldots, x_n) = P(x_2, \ldots, x_n) .$$ (3.3)

What this means is that in terms of the random variables

$$\sum_{x_1} Pr\{X_1 = x_1, X_2 = x_2, \ldots, X_n = x_n\} = Pr\{X_1 = x_2, \ldots, X_{n-1} = x_n\} ,$$

which is just the statement of shift invariance. Finally, a process is iid, if

$$P(x_1, \ldots, x_n) = \prod_i P(x_i) .$$ (3.4)

3.2 Entropy of Stationary Processes

There are two ways to define the entropy of a random process:

1. $H_1(X) = \lim \frac{1}{n} H(X_1, \ldots, X_n)$
2. $H_2(X) = \lim H(X_n|X_{n-1}, \ldots, X_1)$

provided, of course, that the limits exist.

Theorem 13 *For stationary processes $H_1(X) = H_2(X)$.*

Proof. By Equation (9) in section 2.2,

$$H(X_{n+1}|X_n, \ldots, X_1) \le H(X_{n+1}|X_n, \ldots, X_2)$$
$$= H(X_n|X_{n-1}, \ldots, X_1),$$

where the equality follows from stationarity. Hence, $H_2(X)$ as the limit of a non-increasing sequence of positive numbers exists.

Consider then

$$\frac{1}{n} H(X_1, \ldots, X_n) = \frac{1}{n} \sum_{i=1}^{n} H(X_i|X_{i-1}, \ldots, X_1)$$ (3.5)

$$= H_2(X) + \frac{1}{n} \sum_i [H(X_i|X_{i-1}, \ldots, X_1) - H_2(X)] .$$ (3.6)

The difference within the brackets is not greater than ϵ for $i \geq N_\epsilon$, which implies that

$$\frac{1}{n}H(X_1,\ldots,X_n) \leq H_2(X) +$$

$$+ \frac{1}{n}\sum_{i=1}^{N_\epsilon}[H(X_i|X_{i-1},\ldots,X_1) - H_2(X)] + \frac{n-N_\epsilon}{n}\epsilon .$$

The second term goes to zero with increasing n and since the last term is less than ϵ, we certainly have

$$H_1(X) \leq H_2(X) + 2\epsilon .$$

Since we can take ϵ as small as we like, $H_1(X) \leq H_2(X)$. To get the opposite inequality, notice that by stationarity and the fact that increasing the amount of conditioning cannot increase the entropy $H(X_i|X_{i-1},\ldots,X_1) - H_2(X) \geq 0$. Hence by Equation (3.6), $\frac{1}{n}H(X_1,\ldots,X_n) \geq H_2(X)$, and so is the limit H_1.

Because of the theorem we write $H_1(X) = H_2(X) = H(X)$.

3.3 Markov Processes

It is clear that to describe a random process in a constructive way, for instance, for the purpose of fitting it to data, we must consider special types of processes, for otherwise it would take an infinite number of conditional probabilities (3.2) to specify them. The most familiar and important subclasses of processes are those of finite memory:

Definition: A process is a *Finite Memory* process, if

$$P(x_{n+1}|x_n, x_{n-1},\ldots,x_1) = P(x_{n+1}|x_n, x_{n-1},\ldots,x_{n-k}) \equiv P(x_{n+1}|x_{n-k}^n)$$

for all n.

We can then write the recursion

$$P(x^{n+1}) = P(x_{n+1}|x_{n-k}^n)P(x^n)$$
$$= P(x_{n+1}|x_{n-k}^n)P(x_n|x_{n-k-1}^{n-1})\ldots P(x_{k+1}|x^k)P(x^k) ,$$

where the segments x_{n-k}^n are called states. Such a process is also called a *Markov* chain of order k. A first-order Markov chain, then, is one satisfying

$$P(x_1,\ldots x_n) = P(x_n|x_{n-1})P(x_1,\ldots,x_{n-1}) . \tag{3.7}$$

If, in particular, the conditional probabilities $P(x_{n+1}|x_{n-k}^k)$ do not change as a function of time, the process is called *time-invariant*, or often by an abuse of language "stationary". Starting at an initial state, the probabilities of the states can be generated by gliding state transformations and the conditional probabilities as follows:

$$F : x_{n-k}^n, x_{n+1} \mapsto x_{n-k+1}^{n+1} \tag{3.8}$$

$$P(x_{n+1-k}^{n+1}) = P(x_{n+1}|x_{n-k}^n)P(x_{n-k}^n) . \tag{3.9}$$

For the first-order time-invariant chain, the symbols x_t, ranging over $\{1, \ldots, m\}$, define the states, and if we collect the state probabilities into a vector $\bar{P}_t = P(x_t = 1), \ldots, P(x_t = m)$ and the conditional probabilities into a matrix $A = \{p_{ij} = P(i|j)\}$, we get

$$\bar{P}_{t+1} = A\bar{P}_t . \tag{3.10}$$

Due to the properties of the matrix A of the state transition probabilities, the probabilities of the states converge to their stationary limits, written as a column vector $\bar{P} = col(P(1), \ldots, P(m))$. They add up to unity and by Equation (3.10) satisfy

$$\bar{P} = A\bar{P} . \tag{3.11}$$

Example

Let for a binary first-order Markov process $p_{11} = 1 - \alpha$, $p_{12} = \beta$, $p_{21} = \alpha$, and $p_{22} = 1 - \beta$. Solving the stationary-state probabilities with the additional requirement $P(1) + P(2) = 1$ gives $P(1) = \frac{\beta}{\alpha+\beta}$ and $P(2) = \frac{\alpha}{\alpha+\beta}$.

For stationary Markov chains

$$H(X) = H(X_2|X_1) = -\sum_{x_1} p(x_1) \sum_{x_2} p(x_2|x_1) \log p(x_2|x_1) .$$

In case of the first-order Markov process in the example we get

$$H(X) = \frac{\beta}{\alpha + \beta} h(\alpha) + \frac{\alpha}{\alpha + \beta} h(\beta) ,$$

where $h(\alpha)$ denotes the binary entropy function evaluated at α.

3.4 Tree Machines

There are two characteristics of particular importance in Markov chains: the order and the number of free parameters needed to define the process. A Markov chain of order k with an alphabet size m has m^k states, each having $m - 1$ probability parameters. However, there are important applications where the number of states is very large but the number of parameters required to specify the process is small, which makes their estimation relatively easy. Hence, it is worthwhile to separate the two characteristics, which can be achieved by *tree machines*, introduced in [53] and analyzed and developed further in [83]. We describe only binary tree machines.

First, take a complete binary tree, which is characterized by the property that each node has either two successors or none if the node is a leaf. At each node s a conditional probability $P(i = 0|s)$ is stored, which at the root node, λ, is written as $P(i = 0)$. In addition, there is defined a permutation of the indices, which we write as $\sigma(1, 2, \ldots, n) = \tau_1, \tau_2, \ldots, \tau_n$. The intent with this is to order the past symbols $x_n, x_{n-1}, \ldots, x_1$ by their importance in influencing the "next" symbol x_{n+1}, or at least what we think their importance is. Hence, for example, in Markov chains we implicitly assume that the most important symbol is x_n, the next most important x_{n-1} and so on, which is achieved by picking the permutation as the one that reverses the order. In modeling image data we often pick the order as the previous symbol x_n, the symbol x_{n-M} right above the symbol x_{n+1} and so on by the geometric nearness in two dimensions to the symbol x_{n+1}.

Let $s_t = x_{\tau_1}, x_{\tau_2}, \ldots$ be the deepest node in the tree, defined by the past permuted symbols. The tree defines the joint probability

$$P(x_1, \ldots, x_n) = \prod_{t=0}^{n-1} P(x_{t+1}|s_t) \, ,$$

to any string, where we take $s_0 = \lambda$ and s_t the node determined by x^t. Since the tree need not be balanced, the depth of the leaves need not be the same, and the number of parameters can be much smaller than the number of states. For instance, for image data the maximal depth of the leaves could be just three or four, while the number of states would be $O(2^M)$ for $M = 1,000$. Notice that the set of all leaf nodes may not define the states. Here is a simple example showing this. Take the four-leaf tree consisting of the leaves 1, 00, 010, and 011. If we consider the previous symbol $x_n = 1$ as a state, then the next symbol $x_{n+1} = 0$ moves the machine to the state 10, which is not a leaf. Hence, in order to implement a tree machine a sufficient number of past symbols must be stored as states, but still only the parameters in the leaves are needed to compute the probabilities of strings.

We next find the maximum-likelihood estimates of the conditional probabilities $P(0|s)$ at the leaves. We have

$$\log 1/P(x^n) = C(t_0) + \sum_s n_s \sum_{i=0}^{1} n_{i|s} \log 1/P(i|s) \, ,$$

where $C(t_0)$ denotes the initial part in the sum determined by the internal nodes, n_s the number of times leaf s occurs, and $n_{i|s}$ the number of times out of these occurrences symbol i occurs. By the noiseless coding theorem

$$\log 1/P(x^n; \{P(0|s)\}) = C(t_0) + \sum_s n_s \sum_{i=0}^{1} n_{i|s} \log 1/P(i|s) \geq C(t_0)$$

$$+ \sum_s n_s \sum_{i=0}^{1} n_{i|s} \log \frac{n_s}{n_{i|s}} ,$$

where the equality takes place for $P_{i|s} = \frac{n_{i|s}}{n_s}$.

3.5 Tunstall's Algorithm

In the first section we discussed traditional coding, where the data sequences are modeled as samples from a random variable, which when extended to sequences amounts to an iid process. The prefix code with minimum mean length can be found with Huffman's algorithm, which is described in most textbooks on coding and which we discussed briefly above. Our main interest, however, is in codes and models of random processes, which requires a generalization of the traditional codes and even creation of completely different codes. We begin our discussion with a traditional algorithm due to Tunstall, which not only is more suitable for coding of random processes than Huffman's algorithm but which also is behind the important universal coding algorithm due to Lempel and Ziv.

Tunstall's algorithm constructs a binary code tree with a desired number of codewords, say, m, by the rule, Split the node with the maximum probability until m leaves have been generated. The segments of the source strings to be encoded are the paths to the leaves, each of which can be encoded with $\lceil \log m \rceil$ bits – namely, as the binary ordinal of the leaf when they are sorted from left to right. Such a coding is called "Variable-to-Fixed" length coding; the rule of splitting the node with maximum probability clearly tends to equalize the leaf probabilities, which then can be encoded efficiently by equal-length codewords.

A thorough analysis of Tunstall's code is done in [16]. We give here a brief analysis with useful results on trees. We begin with a most convenient formula for the mean length of a finite complete tree, where the node probabilities satisfy the axioms of a random process; i.e., the root has the probability unity and each internal node's probability is the sum of the children's probabilities. Let T_i be such a tree with i leaves $S_i = \{s_1, \ldots, s_i\}$. Its mean length is defined to be

$$L(T_i) = \sum_{s \in S_i} P(s)|s| ,$$

where $|s|$ is the depth of node s. If we fuse any pair of leaves with common father into the father node, say, w, and denote the resulting complete subtree by T_{i-1}, we see that

$$L(T_i) = L(T_{i-1}) + P(w) = \sum_{w \in Int} P(w) , \qquad (3.12)$$

where w runs through all the internal nodes of the tree T_i. To simplify the subsequent discussion we consider only Bernoulli sources. Some of the main results hold for other types of dependent sources as well, although a satisfactory generalization of Tunstall's algorithm to Markov sources is difficult; see [73].

Theorem 14 *Tunstall's algorithm produces the tree with m leaves of maximum mean length.*

Proof. Let T_m^* be a tree with m leaves which has maximum mean length $L(T_m^*)$. The two-leaf tree has length 1, and a three-leaf Tunstall tree is clearly optimal. Arguing by induction, we see in (3.12) that for T_i to be optimal we must split the leaf of T_{i-1}^* that has the maximum probability, which makes T_i optimal among the i-leaf trees. This completes the proof.

A very fast way to encode the leaf segments in a Tunstall tree is to enumerate them from left to right and encode a leaf w by its ordinal, written in binary with $\lceil \log m \rceil$ bits. This Variable-to-Fixed coding is also efficient. First, by the theorem just proved, the ratio

$$r(m) = \frac{\lceil \log m \rceil}{L(T_m)}$$

is minimized by the Tunstall tree over all trees of the same size. Secondly, by the noiseless coding theorem the mean ideal code length for the leaves is the leaf entropy $H(T_m)$. Clearly $\log m$ is an upper bound for the entropy, and the excess of $r(m)$ over the absolute minimum $H(T_m)/L(T_m)$ depends on how much smaller the entropy is than the uniform $\log m$. But again, no tree with m leaves has greater leaf entropy than a Tunstall tree. To see this, consider a Bernoulli process with the parameter $P(0) = p$, which we take to be not smaller than $q = 1 - p$. The entropy defined by the leaf probabilities of the complete Tunstall subtree T_i is given by

$$H(T_i) = -\sum_{s \in S_i} P(s) \log P(s) .$$

Consider again the subtree T_{i-1}, obtained by fusing two leaf sons, say, $w0$ and $w1$, to their father node w. Clearly,

$$\begin{aligned} H(T_i) &= H(T_{i-1}) + P(w) \log P(w) - P(w)p \log(P(w)p) - P(w)q \log(P(w)q) \\ &= H(T_{i-1}) + P(w)h(p) \\ &= h(p) \sum_{w \in Int} P(w) = h(p)L(T_i) , \end{aligned} \qquad (3.13)$$

where $h(p)$ denotes the binary entropy function. This result generalizes to Markov sources where an alphabet extension tree is placed at each state [73].

Now, let $\hat{P}_m$ and $\tilde{P}_m$, respectively, be the maximum and the minimum leaf probabilities of a Tunstall tree with m leaves. Further, for $m = 2$ we have

$\tilde{P}_m \geq q\hat{P}_m$. Then in terms of code lengths, which may be easier to argue with, $-\log \tilde{P}_m \leq -\log \hat{P}_m + \log 1/q$ for $m = 2$, which forms the induction base. We show by induction that the same inequality holds for all larger Tunstall trees. In fact, suppose it holds up to size n, $n \geq m$. Then by splitting the node with $\hat{P}_n$ we create two new nodes with probabilities $p\hat{P}_n$ and $q\hat{P}_n$, the latter of which will be the smallest $\tilde{P}_{n+1}$ in the $n + 1$-leaf tree. By the induction assumption $-\log \tilde{P}_{n+1} \leq -\log P_n(s) + \log 1/q$ for all leaves s in T_n, and we need to check the condition for the new leaf with $p\hat{P}_n$. Indeed, $q\hat{P}_n = \tilde{P}_{n+1} \geq pq\hat{P}_n$, which completes the induction.

Clearly, for all n,

$$\log n \leq \log 1/\tilde{P}_n \leq \log 1/\hat{P}_n + \log 1/q ,$$

while $H(T_n) \geq \log 1/\hat{P}_n$, and $H(T_n) \leq \log n$, which give

$$H(T_n) \leq \log n \leq H(T_n) + \log \frac{1}{q}.$$

All told, we see with (3.13) that

$$r(n) \to h(p)$$

as $n \to \infty$.

3.6 Arithmetic Codes

We saw in the noiseless coding theorem that an optimal prefix code mimics the data-generating random variable. This relationship is particularly clear in a different kind of code, the *Arithmetic Code*, introduced in [50], including the name, and developed further in [44] and [49] and later implemented by many others. The code amounts to a random process, and it is very well suited for coding data modeled by all kinds of random processes. Its outstanding advantage over the concatenation codes is that it needs no alphabet extension. Indeed, to encode the binary strings occurring at every state of a Markov process by a concatenation code we would have to construct an alphabet extension of symbols that do not occur consecutively in the data string. Although this can be done, it would require very complex bookkeeping to keep track of the symbols that occur consecutively at each state. By contrast, each binary symbol can be encoded by an arithmetic code as it occurs just by knowing the probability of its occurrence, which, moreover, may vary in time. No special algorithm for designing the code as in Huffman or Tunstall codes is needed. In fact, since the coding process is so easy, it could be separated from the much more difficult task of finding a good model of the data. Hence there is no need to invent various data compression algorithms in which the coding and modeling parts are mixed together. There are many descriptions of arithmetic codes in textbooks, but here we describe the original.

An arithmetic code is based on two ideas:

1. the code is a cumulative probability on strings of length n, sorted alphabetically, and computed recursively in n. and
2. for practical implementation all calculations are to be done in fixed-size registers.

The first idea has been credited to Shannon, presumably without recursive calculation of the cumulative probabilities, but the solution to the crucial second problem is the key and the foundation of arithmetic coding. We consider first the case where no restriction is placed on the precision, and for the sake of simplicity we consider first the binary alphabet. Letting $P(x^n)$ denote a probability function that defines a random process we take the cumulative probability $C(x^n)$ as the code of x^n. From a binary tree with $s0$, the left son of s, we immediately get the recursion:

$$C(x^n 0) = C(x^n) \tag{3.14}$$
$$C(x^n 1) = C(x^n) + P(x^n 0) \tag{3.15}$$
$$P(x^n i) = P(x^n) P(i|x^n), \ i = 0, 1 . \tag{3.16}$$

A minor drawback of this code is that it is one-to-one only for strings that end in 1. The main drawback is that even if we begin with conditional probabilities $P(i|x^n)$, written in a finite precision, the multiplication in the third equation will keep on increasing the precision, and soon enough we exceed any register we pick. What is needed is to satisfy the marginality condition in a random process without increasing the precision. There are several ways to do it. The most straightforward way is to replace the exact multiplication in the third equation above by the following:

$$\bar{P}(x^n i) = \lfloor \bar{P}(x^n) \bar{P}(i|x^n) \rfloor_q, \ i = 0, 1 ,$$

where $\lfloor z \rfloor_q$ denotes the truncation of a fractional binary number $z = 0.0 \ldots 01 \ldots$ such that there are just q digits following the first occurrence of 1. Notice that the number of the initial zeros grows, since the probability of the string gets smaller and smaller. It is clear then that the addition in the update of the code string can also be done in a register of width $q + 1$, except for a possible overflow, which is dealt with later. We shall show presently that the code is one-to-one on strings that do not end in a zero, and hence decoding can be done.

Consider the two strings x^n and y^m which differ for the first symbol $x_t = 1$ and $y_t = 0$: $x^n = u1w$ and $y^m = u0v$. Unless empty, the strings w and v end in the symbol 1. In the nonempty case the code strings can be written as

$$C(x^n) = C(u) + \bar{P}(u0) + \bar{P}(u10^r 0) + \ldots$$
$$C(y^m) = C(u) + \bar{P}(u00^s 0) + \ldots ,$$

where 0^r and 0^s denote strings of 0 of length $r \geq 0$ and $s \geq 0$, respectively.

We see that regardless of what w and v are,

$$\bar{P}(u0) \leq C(x^n) - C(u), \text{ for } x_t = 1,$$
$$\bar{P}(u0) > C(y^m) - C(u), \text{ for } x_t = 0.$$

Obviously the first is true and so is the second if v is empty. But even when it is nonempty, we have

$$\bar{P}(u0) \geq \sum_i \bar{P}(u0i) > \bar{P}(u00) + \bar{P}(u010) + \bar{P}(u0110) + \cdots ,$$

where the right-hand side results from $v = 11\ldots1$ which gives the largest possible sum of the addends.

We conclude that decoding can be done with the rule

$$x_t = 1 \iff \bar{P}(u0) \leq C(x^n) - C(u) ,$$

which, in turn, can be decided by looking at the $q+1$ leading bits of the code string.

The problem of a possible overflow can be solved by "bit stuffing". In fact, when the register where the code is being updated becomes full, a further addition will propagate a "1" into the already decoded portion of the string. In order to prevent this from happening, an additional bit 0 is added to the code string in the bit position immediately to the left of the register. An overflow will turn this into a "1," which can be detected by the decoder and taken into account.

We make an estimate of the code length. We see in (3.16) that the code length is given in essence by the number of leading zeros in $\bar{P}(x^n0)$ plus the size of the register q needed to write down the probability itself. If the data have been generated by a process $P(x^n)$, we have $\bar{P}(i)/P(i) \geq 1 - 2^{r-q}$, where r is the smallest integer such that $\min_i P(i) \geq 2^{-r}$, and q is taken larger than r. We then have

$$-\log \bar{P}(x^n) \leq -\log P(x^n) + n \log(1 - 2^{r-q}) .$$

Since further also $\bar{P}(x^ni)/\bar{P}(x^n)\bar{P}(i) \geq 1 - 2^{r-q}$, we get

$$\frac{1}{n}L(x^n) \leq -\frac{1}{n}\log P(x^n) + q/n + 2^{r-q} ,$$

which for large enough n can be made as close to the ideal code length as desired by taking q large enough. In this estimate we have excluded the effect of the bit stuffing, which in general extends the code length by an ignorable amount.

The quantity $\bar{P}(x^n)$ no longer defines a probability measure and a random process; rather, it defines a semi-measure. It still requires a multiplication, which sometimes is undesirable. It is possible, however, to remedy these defects with a tolerable increase in the code length. Consider the following recursive definition of a process $p(x^n)$: First, let $p(0)$ denote a probability of the (binary) symbol is being 0, written as a binary fraction with w fractional digits. We take it to be less than or equal to $\frac{1}{2}$; this low probability symbol could of course be 1, and in fact in general it is sometimes 0 or sometimes 1 depending on the past string. Now define $a(x^n) = 2^{L(x^n)}p(x^n)$, where $L(x^n)$ is the integer that satisfies $.75 \le a(x^n) < 1.5$. In other words, for a string $s = x^n$, $a(s)$ is the decreasing probability $p(s)$ normalized to a fixed-size register. Then calculate recursively

1. $p(s0) = 2^{-L(s)}p(0)$, and
2. $p(s1) = p(s) - 2^{-L(s)}p(0)$.

We see that $p(s)$ has no more than w digits following the first 1 and that it defines a random process.

Encoding Algorithm

Initialize register C with 0's and set A to 10...0.

1. Read next symbol. If none exists, shift contents of C left w positions to obtain the final code string.
2. If the next symbol is the low-probability symbol, say, x, replace the contents of A by $p(x)$ $(p(x) \le 1/2)$ and go to 4.
3. If the next symbol is the high probability symbol, add to register C the number $p(x)$ and subtract from A the same number.
4. Shift the contents of both registers left as many positions as required to bring A to the range $[.75, 1.5)$. Go to 1.

Decoding Algorithm

Initialize register C with w leftmost symbols of the code string and set A to 10...0.

1. Form an auxiliary quantity T by subtracting $p(x)$ from the contents of C. Test if $T < 0$.
2. If $T < 0$, decode the symbol as x. Load A with $p(x)$.
3. If $T \ge 0$, decode the symbol as the high-probability symbol. Load C with T, subtract $p(x)$ from A.
4. Shift the contents of both registers left as many positions as required to bring A to the range $[.75, 1.5)$ and read the same number of symbols from the code string in the vacated positions in C. If none exist, stop. Otherwise, go to 1.

We now give the formulas for the non-binary alphabet, say, $A = \{0, 1, \ldots, m-1\}$. Denoting the conditional probabilities by $P(i|x^n)$, we get from a balanced m'ary tree of depth $n+1$ with si the i-th son leaf of s the recursion:

$$C(x^n 0) = C(x^n) \tag{3.17}$$

$$C(x^n i) = C(x^n) + P(x^n) \sum_{j<i} P(j|x^n) \tag{3.18}$$

$$P(x^n i) = P(x^n) P(i|x^n). \tag{3.19}$$

As above, we then replace both the conditional and the string probabilities by their quantizations

$$\bar{P}(i|x^n) = \lfloor P(i|x^n) \rfloor_q \tag{3.20}$$

$$\bar{P}(x^n i) = \lfloor \bar{P}(x^n) \bar{P}(i|x^n) \rfloor_q, \ i \in A, \tag{3.21}$$

where $\lfloor z \rfloor_q$ denotes the truncation of a fractional binary number $z = 0.0\ldots$ $01\ldots$ such that there are just q digits following the first occurrence of 1. The coding operations are then given by

$$C(x^n 0) = C(x^n) \tag{3.22}$$

$$C(x^n i) = C(x^n) + \bar{P}(x^n) \sum_{j<i} \bar{P}(j|x^n), \ i > 0 \tag{3.23}$$

together with the recursion above for $\bar{P}(x^n i)$.

We mention in closing that any concatenation code can be represented as an arithmetic code. This is because if we design an arithmetic code from the probabilities $2^{-\ell_i}$, where ℓ_i is the code length of the symbol i in a prefix concatenation code, the concatenation operation becomes a shift followed by an addition, and the multiplication in the probability recursion becomes simply a shift. We illustrate this with the example of the Huffman code above, which gave the codewords 00,01,10,110,111. Write the alphabet $A = \{a, b, c, d, e\}$. Pick the symbol probabilities as $P(a) = P(b) = P(c) = 2^{-2}$ and $P(d) = P(e) = 2^{-3}$. The code string is a fractional binary number, and we represent $C(a)$ as .00, and tack this at the end of the code string by (3.17) thus: $C(x^n a) = C(x^n)00$, which as a number equals $C(x^n)$. Similarly, we retain the trailing 0's on any codeword that ends in 0. Then $C(a) = .00$, $C(b) = .01$, $C(c) = .10$ and so on. Continuing by (3.18) and (3.19) we get

$$C(bc) = .01 + .01 \times (.01 + .01) = .01 + .0010 = .0110$$
$$P(bc) = .01 \times .01 = .0001 \ .$$

We see that the concatenation operations are mimicked exactly.

For an interpretation of arithmetic codes as number representations we refer to [52].

3.7 Universal Coding

In practice the data compression problem is to encode a given data string when no probability distributions are given. For a long time the problem was tackled by designing codes on an intuitive basis, until it was realized that such codes always incorporate some type of a model which is "universal" in an entire class $\mathcal{M} = \{P_i(x^n)\}$ of stationary processes, where i ranges over a countable set Ω. Since we know by the noiseless coding theorem how to design a code with each member in the class so that the mean length is close to the corresponding entropy, a reasonable requirement of any code which claims to represent the entire class and be *universal*, relative to the class, is that it has the following property:

The mean per symbol code length should approach the per symbol entropy in the limit, no matter which process in the class generates the data:

$$\frac{1}{n}E_iL(X^n) \to H_i(X) , \qquad (3.24)$$

where the expectation is with respect to P_i and $H_i(X)$ is the corresponding per symbol entropy.

If we can construct or even define codes with such a property, we may start asking more: We would like the mean per symbol length to approach the entropy at the fastest possible rate. A little reflection will reveal that this is too much to ask, for by the noiseless coding theorem we could achieve the entropy instantly if we were to guess the data-generating process correctly. However, if there are a countable number of index values we cannot expect to be right with our guess, and the exceptional case does not seem to be of much interest in practice. What we are interested in is to find the maximum rate at which the entropy is reached without knowing or guessing the right data-generating process.

3.7.1 Lempel–Ziv and Ma algorithms

An elegant algorithm for universal coding is due to Lempel and Ziv [37]. The algorithm parses the data string x^n recursively into nonoverlapping segments by the rule –

- Starting with the empty segment, each new one added to the collection is one symbol longer than the longest match so far found.

Lempel and Ziv encoded each parsed segment as the pair (i, y), where i, written as a binary number, gives the index of the longest earlier found match in the list, and y is the last added symbol. The code length for the string x is then given by

$$L_{LZ}(x) = m(x) + \sum_{j=1}^{m(x)} \lceil \log j ,$$

where $m(x)$ denotes the number of parsed segments.

It may appear that there is no probability model behind the algorithm. However, parsing actually incorporates occurrence counts of symbols, which in effect define a random process and a sequence of Tunstall trees permitting a Variable-to-Fixed length coding (more or less, because the tree keeps on growing!). To see this, consider the recursive construction of a binary tree from a given binary string $x_1, \ldots, x_n$:

1. Start with a 2-node tree, the root node marked with a counter initialized to 2, and each son's counter initialized to 1.
2. Recursively, use the previous tree to parse the next segment off the remaining string as the path from the root to the deepest node in the so far constructed tree. While climbing the tree, increment by one the count of each node visited.
3. When the last visited node is reached, split it to create two new son nodes, and initialize their counts to 1.

If we denote the parsed segments by σ_i, we can write the string in these terms as: $x^n = \sigma_1, \sigma_2, \ldots, \sigma_m$, which corresponds exactly to the parses found with the LZ algorithm; the last segment σ_m may not be complete. Notice that the leaves of the tree have the count of unity, and each father node's count is the sum of the sons' counts. Hence, if we divide all the node counts $n(s)$ by the root count $n(\lambda)$, each node s gets marked with the probability $P(s) = n(s)/n(\lambda)$. Also, the leaf with the highest count, which will be 2 before the split, gets split – just as in Tunstall's algorithm.

Write $x^t = \sigma_1 \ldots \sigma_i z_j$, where z_j denotes a prefix of the full segment σ_{i+1}. By identifying z_j with the corresponding node in the so far constructed tree we can define the conditional probability of the symbol x_{t+1}, given the past string, as

$$P(x_{t+1}|x^t) = n(z_j x_{t+1})/n(z_j) .$$

When we multiply all these conditional probabilities along the next fully parsed segment σ_{i+1} we get its conditional probability as

$$P(\sigma_{i+1}|\sigma^i) = 1/n(\lambda) ,$$

where $n(\lambda) = i + 2$. This last equality follows from the fact that each parsed segment adds 1 to the root count, which is initialized to 2. Finally, the probability of a string x with $m(x)$ full segments is

$$P(x) = 1/(m(x) + 1)! .$$

This gives the *ideal* code length as $- \log P(x)$, which is somewhat better than the Lempel–Ziv code length, because of the fairly crude way the coding is done in that code.

The Lempel–Ziv (LZ) code is universal in a very large class of processes $\{P\}$, namely, the class of stationary ergodic processes, in the sense that

$$\frac{1}{n} E_P L_{LZ}(X^n) \to H(P) ,$$

no matter which process P generates the data. The notation E_P denotes the expectation with respect to the process P. The same holds also in the almost sure sense. Hence, it does satisfy the minimum optimality requirement stated above. The rate of convergence can be shown to be

$$\frac{1}{n}[E_\theta L(X^n) - H(X^n)] = O\left(\frac{1}{\log n}\right),$$

which in the case of the class of Markov and tree machine processes $P(x^n; T)$ will be seen not to be too good.

Although Ma's version of the LZ algorithm does define a random process, which gives a universal model for the huge class of stationary ergodic processes, it is somewhat odd in that the probability of the next symbol given the past, $P(x_{t+1}|x^t)$, depends greatly on where in the last parsed segment the previous symbol x_t lies. For instance, if x_{t+1} is the very last symbol of a segment, its conditional probability is $\frac{1}{2}$, which is as bad as it can be. For this reason we consider the LZ algorithm to give a good universal model for data compression but not for statistical modeling.

The universal tree machine discussed in Applications gives a superb model for the class of Markov chains, which is good for diverse applications of models. In fact, even for data compression, the best universal coding systems are modifications of the algorithm such that they can handle symbols that have not yet occurred in a state or context. This problem is important in all context trees, especially when they are grown predictively, because no symbols are seen initially. The same applies even when a tree is applied to new data.

Historical Notes

My inspiration for arithmetic codes was the *combinatorial* information presented in the elegant short paper by Kolmogorov [34]. I applied it immediately to binary sequences by developing a recursive enumeration scheme involving factorials, which raised the problem of how to approximate them in fixed-size registers. With some struggle I solved the problem, and the so-called LIFO (Last-In First-Out) arithmetic code was born [50]. The quantization problem turned out to be much easier for the FIFO (First-In First-Out) codes, as shown by Pasco in his doctoral thesis [44], who was aware of my work. Enumerative coding without recursion or quantization is discussed in [11].

When arithmetic coding was invented, IBM applied and got a patent for it. The research management's failure to see the value of what the company had kept the codes more or less dormant until the patent protection ended in the nineties. After that its applications exploded, and currently it has virtually replaced the concatenation codes.

In the early universal coding algorithms there was no clear separation of the model part from the coding. Arithmetic coding made such a separation possible, although even today there is some confusion, and arithmetic coding is sometimes regarded as a universal data compression system.

Part II

Statistical Modeling

Statistical modeling is about finding general laws from observed data, which amounts to extracting information from the data. Despite the creation of information theory half a century ago with its formal measures of information, entropy, and the Kullback–Leibler distance or the relative entropy, there have been serious difficulties in applying them to make exact the idea of information extraction for model building. It is perhaps an indication of the strength of dogma in statistical thinking that only relatively recently (in 1973) was the information extraction process formalized by Akaike [1], as one of searching for the model in a proposed collection that is closest to the "true model" as a probability distribution in terms of the Kullback–Leibler distance. The "true model" is assumed to lie outside the collection, and it is known only through the observed data from which the distance must be estimated. This turns out to be difficult, and Akaike's criterion AIC amounts to an asymptotic estimate of the mean Kullback–Leibler distance, the mean taken over the estimated models in the class, each having the same number of parameters.

Although Akaike's work represents an important deviation in thinking from tradition in avoiding the need to add artificial terms to the criterion to penalize the model complexity, there are difficulties with the criterion. The first, of course, is the need to assume the existence of the "true" underlying model. As a matter of fact, even in such a paradigm the Kullback–Leibler distance gets reduced as the number of parameters in the fitted models is increased, which is the very problem we are trying to avoid. That this gets overcome by the estimated mean distance is not a great consolation for how can we explain the paradox that the ideal criterion, where everything is known, fails, but the estimation process, which one would expect to make things worse, produces a criterion that works better than the ideal. Besides, it is known that if we assume the "true" model to be in the set of the fitted models, the AIC criterion will not find it no matter how large the data set is, which suggests a lack of self-consistency. Even though the derivation of Akaike's criterion is shaky, and due to the lack of clear data-dependent interpretation it must be considered ad hoc, it seems to work reasonably well, especially in cases where the data cannot be well modeled by a parametric model. This is simply an empirical experience for which no explanation appears to have been made.

There is another quite different way to formalize the problem of extracting information from data, which appears to be much more natural. It is based on the idea, inspired by the theory of Kolmogorov complexity which will be reviewed below, that the complexity of a data set is measured by the fewest number of bits with which it can be encoded when advantage is taken of a proposed class of models. Hence, the complexity measure is relative to the class of models, which then act as a language allowing us to express the properties in the data, and, as we shall see, the information in the data. This makes intuitive sense, for if the language is poor, we expect to be able to learn only gross properties. If, on the other hand, the language is very rich, we can express a large number of properties, including spurious "random" quirks. This raises the thorny issue of deciding how much of and which properties of the data we want to and can learn. Our solution will be based on the idea that the portion in the data that cannot be compressed with the class of models available will be *defined* to be uninteresting "noise", and the rest is what we want to learn – the useful learnable *information*. We may state that to achieve such a decomposition of data is the purpose of all modeling.

When we formalize the two fundamental notions, the complexity and information, relative to a class of models, there is no need to assume any metaphysical "true model", which somehow would represent not only the information in the given data set but even in all future data sets generated by sampling the "true model". Although our formalization does provide a solid foundation for a theory of modeling, it of course does not solve all the modeling problems, for the selection of a good class of models remains. In fact, there can be a lot of prior information that others have gathered from data generated under similar conditions. However, this we cannot formalize in any other way than by saying that such information should be used to suggest good or at least promising classes of models. The noncomputability of the Kolmogorov complexity (see below) implies that the process of selecting the optimal model and model class will always have to be done by informal means where human intelligence and intuition will play a dominant role. No other currently used methodology of model selection has revealed similar limits to what can and cannot be done.

4

Kolmogorov Complexity

For a string (x^n), generated by sampling a probability distribution $P(x^n)$, we have already suggested the ideal code length $-\log P(x^n)$ to serve as its complexity, the Shannon complexity, with the justification that its mean is for large alphabets a tight lower bound for the mean prefix code length. The problem, of course, arises that this measure of complexity depends very strongly on the distribution P, which in the cases of interest to us is not given. Nevertheless, we feel intuitively that a measure of complexity ought to be linked with the ease of its description. For instance, consider the following three types of data strings of length $n = 20$, where the length actually ought to be taken large to make our point:

1. 01010101010101010101
2. 00100010000000010000
3. generate a string by flipping a coin 20 times

We would definitely regard the first string as "simple", because there is an easy rule permitting a short description. Indeed, we can describe the string by telling its length, which takes about $\log n$ bits, and giving the rule. If the length n were actually much greater than 20, the description length of the rule "alternate the symbols n times starting with 0," encoded in some computer language as a binary string, could be ignored in comparison with $\log n$. The amount of information we can extract from the rule is also small, because there is not much information to be extracted.

The second string appears more complex, because we cannot quite figure out a good rule. However, it has $n_0 = 17$ zeros and only $n - n_0 = 3$ ones, which may be taken advantage of. In fact, there are just $N = 20!/(3!17!)$ such strings, which is quite a bit less than the number of all strings of length 20. Hence, if we just sort all such strings alphabetically and encode each string as its ordinal, written in binary, we can encode the given string with about $\log N \cong 14$ bits. This suggests that we are modeling the string more or less by the class of Bernoulli models. The information that can be extracted that way is a bit more than in the previous case, but the amount is not all that much, just what it takes to encode the ratio 3/20.

The third string is maximally complex; no rule helps to shorten the code length, about $n = 20$, which results if we write it down symbol for symbol. However, the amount of information is small; after all, there is nothing to be learned other than that the string looks random.

4.1 Elements of Recursive Function Theory

An excellent introduction to the recursive function theory is [41], and a more thorough treatment is [38]. The objective in the recursive function theory is to consider functions, first from the set of natural numbers $\mathcal{N}$ to itself, or equivalently, from the set of finite binary strings B^* to itself, and more generally between cartesian products of both of these sets, which can be defined by *effective* operations (instructions) like computer programs. There are two different styles of presentation, one purely mathematical and the other in terms of computer programs. The latter can be intuitive but it has its dangers, for a rigorous proof of a theorem amounts to a detailed description of a computer program. This is often long and when its description is abbreviated it is a matter of one's experience in computer jargon if the proof can be understood. We give mostly only the important definitions and begin with *primitive* and *partial recursive functions (prf)* for the natural numbers and their pairs, triples, and so on as follows:

1. constant functions
2. successor function $S(n) = n + 1$ and
3. projection $x^k = (x_1, \ldots, x_k) \mapsto x_i$ are primitive recursive. So are
4. compositions: $g_1, \ldots, g_k$ and $f(x_1, \ldots, x_k)$ primitive recursive functions imply that $f(g_1, \ldots, g_k)$ is primitive recursive. Further
5. f and g primitive recursive imply that

$$h(0, x^k) = f(x^k)$$
$$h(n + 1, x^k) = g(h(n, x^k), n, x^k)$$

   is primitive recursive.
6. Primitive recursive functions are prf. If $f(t, x_2, \ldots, x_k)$ is prf then

$$g(x_1, x_2, \ldots, x_k) = \min_t[f(t, x_2, \ldots, x_k) = x_1]$$

   is prf.

The last condition is the one that generalizes the primitive recursive functions to partial recursive functions and forms a closed set of functions of crucial importance. Here is an example of its use

$$\sqrt{n} = \min_t[t^2 = n].$$

Notice that $\sqrt{n}$ as an integer is undefined for all integers.

A basic result in the theory states that the set of prf functions is the same as the set of functions, where the calculation $n \mapsto f(n)$, if it exists, can be described by a computer program such as a Touring machine.

A prf function f is *total*, or *recursive*, if it is defined for all of its arguments. A total prf is not necessarily a primitive recursive function.

The range and the domain of the prf functions are generalized as follows. A real number is *computable* if there is a single program to compute each of its digits. The requirement of a single program is crucial. It is clear that if a different program is permitted to calculate each digit any real number would be computable. Hence, for instance π is computable. Notice, too, that the set of computable real numbers is countable, because the set of all computer programs is countable.

Similarly, a function is computable if its values are computable. These are further generalized as follows. A number or function is semicomputable from above (below) if there is a program to calculate converging approximations of it from above (below). However, we may never know how far from the limit the approximations are. A function is computable if it is semicomputable both from above and below, because now the approximation accuracy is known.

We conclude this section by deriving a simple upper bound for the ideal code length for the integers 2.4 and at the same time give a construction for a prefix free set for both finite binary strings and natural numbers. Our presentation follows [79]. We begin with an 1-1 map between the natural numbers $\mathcal{N}$ and the set of finite binary strings B^*, sorted first by length and then for strings of the same length lexically thus

$$\lambda, 0, 1, 00, 01, 10, 11, \ldots.$$

Hence, for instance, the string 11 represents the natural number (ordinal) 7, if we start by assigning to the empty string λ the number 1. Call the map which takes the natural number n to the n'th binary string, $bin(n)$. By appeal to intuition this is primitive recursive.

Next, define the prefix encoding of a binary string $x = x_1, \ldots, x_k$ by

$$\bar{x} = 0x_10x_2\ldots0x_k1,$$

which sets a 0 in front of each binary symbol and 1 to the end to indicate that x_k is the last digit of x. We see that $\bar{\lambda} = 1$, $\bar{0} = 001$, $\bar{1} = 011$, and so on. The length of the string $\bar{x}$ is $|\bar{x}| = 2|x| + 1$. Define a new prefix encoding of binary strings as

$$d(x) = \overline{bin(|x|)}x. \tag{4.1}$$

For instance

$$d(0101) = \overline{bin(4)}0101 = 000010101$$

since the length of 0101 is 4, and the fourth binary string is 00, which the bar-encoding converts into 00001. This is also primitive recursive. The length is

$$|d(x)| = |x| + 2\lfloor \log |x| \rfloor + 1.$$

We have converted B^* into the prefix free set $S = \{d(x) : x \in B^*\}$. The natural numbers are now encoded in a prefix manner as $n \mapsto bin(n) \mapsto d(bin(n))$ of length

$$|d(bin(n))| = |bin(n)| + 2\lfloor \log |bin(n)| \rfloor + 1 \approx \log n + 2 \log \log n,$$

which gives the desired upper bound for 2.4.

The function $bin(n)$ generalizes to the function $pair : n \leftrightarrow (k, m)$: $(1, 1) < (1, 2) < (2, 1) < (3, 1) < (2, 2) < (1, 3) < (1, 4) \dots$, which is also primitive recursive.

A set $A \subset \mathcal{N}$ (or $A \subset B^*$) is *recursive*, if its characteristic function is recursive

$$n \in A \Rightarrow I_A(n) = 1$$
$$n \notin A \Rightarrow I_A(n) = 0.$$

Further, A is *recursively enumerable* (re) if A is empty, or it is the range of some total prf function

$$f : \mathcal{N} \to \mathcal{A}$$

with repetitions allowed. Hence A is re if some Touring machine or computer program can list its elements in some order with repetitions allowed

$$A = \{f(1), f(2), \dots\}$$

A fundamental result in the entire theory is the theorem

Theorem 15 *The set of prf functions $\{f : B^* \to B^*\}$ is re; i.e. for some prf function $F : \mathcal{N} \times B^* \to B^*$ of arity two*

$$F(i, x) = f_i(x)$$

all i, whenever $f_i(x)$ exists.

The function $F(i, x)$ is *universal* in that it can implement any partial recursive function. Also the set of prf functions of arity two is re, and so on, and a universal prf function for tuples of strings or computable numbers may be taken as an abstract notion of a universal computer U.

Take a universal computer U and its programs $\{p_1, p_2, \dots\}$ as the language $\mathcal{L}$, sorted first by length and then lexically. The language is re. Also the set of all programs of arity 0; i.e. the input is λ, and $U(p) = x$ for any string x is re by

$$F(i, \lambda) = p_i(\lambda) = x.$$

All the usual programming languages are universal in that any partial recursive function of any number of arguments can be programmed in them. Let U be a computer that can execute programs p, each delivering the desired binary data string $x^n = x_1, \ldots, x_n$ as the output from the empty string λ. When the program p is fed into the computer, the machine after a while prints out x^n and stops. In the coding terminology above a program p is a codeword of length $|p|$ of the string x^n, and we may view the computer as defining a many-to-one map $U : p \mapsto x^n$ which decodes the binary string x^n from any of such programs. The computer can of course execute many other programs, and deliver different types of outputs.

4.2 Complexities

The original idea of the definition of complexity of a finite binary string relative to a universal computer U, which may be called Solomonoff-Kolmogorov-Chaitin complexity of the first type, [9, 34, 71], is defined as:

$$SKC_U(x^n) = \min\{|p| : U(p) = x^n, \, p \in \mathcal{L}\}.$$

This turned out to have a major defect that hampered the development of the theory. The problem is that the language of the computer is not a prefix free set. We know however how to encode a binary string x by a prefix code $d(x)$, and it was Chaitin who added this requirement to the definition. It is still customary to call the result the *Kolmogorov complexity* of a string x^n, relative to a universal computer U:

$$K_U(x^n) = \min\{|d(p)| : U(p) = x^n\}.$$

In words, it is the length of the shortest prefix free program in the language of U that generates the string. A fact of some interest is that the shortest program as a binary string is random in the sense that there is no program that could shorten it. There are other variants of the complexity, discussed extensively in [38]. There are clearly a countable number of programs for each data string, because we can add any number of instructions canceling each other to create the same string.

The set of all programs for U may be ordered, first by the length, and then the programs of the same length alphabetically. Hence, each program p has an index $i(p)$ in this list. Importantly, this set has a generating grammar, which can be programmed in another universal computer's language with the effect that each universal computer can execute another computer's programs by use of this translation or compiler program, as it is also called. Hence, in the list of the programs for U there is somewhere a shortest program p_V, which is capable of translating all the programs of another universal computer V. But this then means that

$$K_V(x^n) \le K_U(x^n) + |p_V| \,,$$

where $|p_V|$ does not depend on the string x^n. By exchanging the roles of U and V we see that the dependence of the complexity of long strings x^n on any particular universal computer used is diluted. We cannot, of course, claim that the dependence has been eliminated, because the constant $|p_V|$ can be as large as we wish by picking a "bad" universal computer. Moreover, it is clear that the shortest program for a string may well have instructions to move data forth and back, which have little to do with the regular features in the string, and hence the syntax is somewhat of a hindrance adding irrelevant bits to the complexity. To get rid of this, we should perhaps define the complexity in an axiomatized recursive function theory. However, by recognizing that for a reasonably long string x^n, its shortest program in a universal computer will have to capture essentially all the regular features in the string, we can safely pick a U, and the Kolmogorov complexity $K_U(x^n)$ provides us with a virtually absolute measure of a long string's complexity.

The Kolmogorov complexity provides a universal model par excellence, which we can even equate with a probability measure, namely,

$$P_K(x^n) = C2^{-K_U(x^n)} \,, \qquad (4.2)$$

where

$$C = 1/ \sum_{y \in B^*} 2^{-K_U(y)} \,,$$

the sum being finite by the Kraft inequality. The universal distribution in (4.2) has the remarkable property of being able to mimic *any* computable probability distribution $Q(x^t)$ in the sense that

$$P_K(x^t) \ge AQ(x^t) \,, \qquad (4.3)$$

where the constant A does not depend on x^t.

4.3 Kolmogorov's Structure Function

Although we have identified the shortest program for a string x, or perhaps the universal probability, with its "ideal" model, because it will have to capture all the regular features in the string, it does not quite correspond to the intuitive idea of a model, especially if the data string has both regular features and a random "noise" like part. In fact, we associate with a model only the regular features rather than the entire string-generating machinery, which is what the shortest program will have to capture. For instance, a purely random string has no regular features, and we would like its "model" to be empty, having no complexity. Hence, we would like to measure separately the complexity of the regular features, or the model, and the rest, which we define to be "noise". Since it is the regular features that we want to learn from data, we call that complexity the *algorithmic information*. In Shannon's theory "information" actually represents the complexity of the messages, and since the distribution as a model of the messages is always taken as given, we have the curious situation that the learnable statistical information there is zero, while the complexity of the messages measures just noise!

We outline next the construction due to Kolmogorov in which the complexities of a model and the "noise", as it were, are separated from the overall complexity of the data; see [79]. But first, we point out that the Kolmogorov complexity is immediately extended to the conditional complexity $K(x|y)$ as the length of the shortest program that generates string x given another string y and causes the computer to stop. One can show [38] that

$$K(x^n, y^n) < K(x^n) + K(y^n|x^n) + O(\log n) . \tag{4.4}$$

We now take y to be a program that describes the 'summarizing properties' of string $x = x^n$. A "property" of data may be formalized as a set A to which the data belong along with other sequences sharing this property. The amount of properties is in inverse relation to the size of the set A. Hence, the smallest singleton set $A = \{x^n\}$ represents *all* the conceivable properties of x^n, while the set consisting of all strings of length n assigns no particular properties to x^n (other than the length n). We may now think of programs consisting of two parts, where the first part describes optimally a set A with the number of bits given by the Kolmogorov complexity $K(A)$, and the second part merely describes x^n in A. The second part should not be measured by the conditional complexity $K(x^n|A)$, because it might take advantage even of properties not in A. Rather, we measure it by $\log |A|$, where $|A|$ denotes the number of elements in A. There are other ways to measure this, some of which are described in [79], and we describe one later when we discuss statistical modeling. The sequence x^n gets then described in $K(A) + \log |A|$ bits.

Consider the so-called *structure function*

$$h_{x^n}(\alpha) = \min_{A \in \mathcal{A}} \{\log |A| : x^n \in A, K(A) \leq \alpha\}. \tag{4.5}$$

Clearly, $\alpha < \alpha'$ implies $h_\alpha(x^n) \geq h_{\alpha'}(x^n)$ so that $h_{x^n}(\alpha)$ is a nonincreasing function of α with the maximum value $n = \log 2^n$ at $\alpha = \log n$, or $\alpha = 0$ if n is given, and the minimum value $h_{x^n}(\alpha) = \log |\{x^n\}| = 0$ at $\alpha = K(x^n)$. In order to avoid pathologies we restrict the sets A to an unspecified class $\mathcal{A}$. After all, we are interested in certain statistical properties of strings, and there are finite sets which incorporate quite weird properties which nullify the entire construct. In our real applications of the idea of structure function later we specify the desired properties.

Because $\log |A| + K(A) \overset{>}{_{\cdot}} K(x^n)$, where $\overset{>}{_{\cdot}}$ denotes inequality to within a constant, (similarly for $\overset{.}{=}$), $h_{x^n}(\alpha)$ is above the line $L(\alpha) = K(x^n) - \alpha$ to within a constant. They are equal up to a constant at the smallest value $\alpha = \bar{\alpha}$ which satisfies

$$\min\{\alpha : h_{x^n}(\alpha) + \alpha \doteq K(x^n)\} , \tag{4.6}$$

and we get the Kolmogorov's *minimal sufficient statistic* decomposition of x^n:

$$\min_{A} h_{x^n}(K(A)) + K(A) , \tag{4.7}$$

which amounts to the *MDL* principle applied to the two-part code length $L(x^n, A) = h_{x^n}(K(A)) + K(A)$. There is a subtle difference, though. Once the properties have been extracted, we want the remaining part to represent the common code length for each of the set of all "noise" sequences. Notice too that the structure function does not talk about "noise" as a sequence; it just gives its amount, measured in terms of code length. The problem of actually separating "noise" as a sequence from the data sequence belongs to "denoising", which we discuss later.

Finally, while there is no point in talking about a "true" model A for the data; i.e., a "true" set A that includes the data, we can talk about the optimal model of complexity $\bar{\alpha}$, as illustrated in Figure 4.1.

All these findings, while very interesting and conceptually important, will not solve the universal coding and modeling problems. This is because the Kolmogorov complexity is noncomputable. What this means is that there is no program such that, when given the data string x^n, the computer will calculate the complexity $K_U(x^n)$ as a binary integer. The usual proof of this

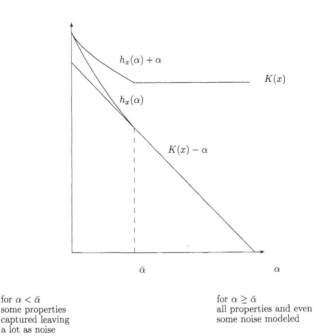

for $\alpha < \bar{\alpha}$
some properties
captured leaving
a lot as noise

for $\alpha \geq \bar{\alpha}$
all properties and even
some noise modeled

Fig. 4.1. A schematic illustration of the structure function and the optimal complexity of the model.

important theorem due to Kolmogorov appeals to the undecidability of the so-called halting problem. Here is another proof given by Ragnar Nohre [43].

Suppose to the contrary that a program q with the said property $q : x^n \mapsto K(x^n)$ exists. We can then write a program p, using q as a subroutine, which finds the shortest string z^n such that

$$K(z^n) > |p| \ .$$

In essence, the program p examines the strings x^t, sorted alphabetically by nondecreasing length, one after another, computes $K(x^t)$ with the subroutine q, and checks if this inequality holds. It is clear that such a shortest string exists, because $K(x^t)$ has no upper bound and $|p|$ has some fixed finite length. But by the definition of Kolmogorov complexity, $K(z^n) \leq |p|$, which contradicts the inequality shown. Hence, no program q with the assumed property exists.

It is possible to estimate the complexity from above with increasing accuracy, but we cannot have any idea of how close to the complexity we are. Despite the negative content of the proved result, it is, or should be, of utmost importance for statistical inference, because it clearly sets a limit for what can be meaningfully asked. Any statistical effort trying to search for the "true" underlying data-generating distribution by systematic (= mechanical) means is hopeless – even if we by the "truth" mean only the best model. Therefore, human intuition and intelligence in this endeavor are indispensable. It is also clear in this light that, while many physical phenomena are simple, because their data admit laws (or, as Einstein put it, God is good), to find the laws is inherently difficult! It has generally taken geniuses to discover even some of the simplest laws in physics. In the algorithmic theory most strings are maximally complex, although we cannot prove a single one to be so, which raises the intriguing question as to whether most data strings arising in the physical world are also maximally complex or nearly so – i.e., obeying no laws. The strings generated in the world's economic systems appear to provide good candidates.

Historical Notes

The idea of Kolmogorov complexity was actually introduced first by Solomonoff [70] and later by Kolmogorov [35] and Chaitin [9]. Kolmogorov's structure function was presented in a talk at the Tallin conference in 1973 (see [12]) which was unpublished; I learned it from P. Vitanyi [79]. The Kolmogorov complexity was initially called "information," and it was criticized by some on the grounds that a purely random binary sequence, which has the maximum amount of information according to the definition, actually has no information to teach us. The answer to this paradoxical state of affairs is provided by the structure function and the associated minimal sufficient statistics notion.

The terminology of calling the Kolmogorov complexity of the regular features in the data the "algorithmic information" is not standard, maybe because of the clash with Shannon's terminology. However, I see its role in modeling as crucial, and rather than calling it the *model cost*, as done frequently in the *MDL* literature, or even worse, the *regret*, in the coding theory, both having clearly a negative connotation, I find the name appropriate. Even the name *model complexity* does not suggest the crucial idea that it measures the amount of the learnable useful information in the data.

5

Stochastic Complexity

As we discussed above, the problem of main interest for us is to obtain a measure of both the complexity and the (useful) information in a data set. As in the algorithmic theory, the complexity is the primary notion, which then allows us to define the more intricate notion of information. Our plan is to define the complexity in terms of the shortest code length when the data is encoded with a class of models as codes. In the previous section we saw that this leads into the noncomputability problem if we let the class of models include the set of all computer programs, a "model" identified with a computer program (code) that generates the given data. However, if we select a smaller class, the noncomputability problem can be avoided, but we have to overcome another difficulty: How are we to define the shortest code length? It seems that in order not to fall back to the Kolmogorov complexity we must spell out exactly how the distributions as models are to be used to restrict the coding operations. Here we adopt a different strategy: we define the idea of shortest code length in a probabilistic sense, which turns out to satisfy all practical requirements – unless the data strings are too short. It is clear that if the strings are short there are too many ways to model them. As a matter of fact, the strings must be long even for the desired algorithmic results to hold.

5.1 Model Classes

For data sets of type $(y^n, x^n) = (y_1, x_1), \ldots, (y_n, x_n)$, where $y_t \in Y$ and $x_t \in X$ are data of any kind, consider a class of models defined by conditional probability or density functions

$$\mathcal{M}_\gamma = \{f(y^n | x^n; \theta, \gamma) : \theta \in \Omega_\gamma \subset R^k, \gamma \in \Gamma\},$$

and the union $\mathcal{M} = \bigcup_\gamma \mathcal{M}_\gamma$. The conditioning makes little difference to the theory, and we consider the models on single data sequences, which we now denote by x^n rather than y^n. The parameter γ is a structure index and $\theta = \theta_1, \ldots, \theta_{k(\gamma)}$ a vector of real-valued, parameters ranging over a subset Ω_γ of the k-dimensional Euclidean space. Depending on whether the data x_t are discrete or real-valued, the models are either probability mass functions of densities. Again the developments are similar.

Example 1. Take the class of ARMA processes

$$x_n + a_1 x_{n-1} + \ldots + a_p x_{n-p} = b_0 e_n + \ldots + b_{q-1} e_{n-q+1} , \qquad (5.1)$$

where the pair (p, q) plays the role of γ and the collection of the coefficients forms θ. These equations define the required probability distribution if we model the process e_n as an iid process, each random variable having the normal distribution of zero mean and unit variance.

5.1.1 Exponential family

An important class of probability models is the exponential family

$$f(x^n; \lambda) = Z^{-1}(\lambda) e^{-\lambda' \phi(x^n)} , \qquad (5.2)$$

where $\lambda = \lambda_1, \ldots, \lambda_k$ denotes a parameter vector and $\phi(x^n) = \phi_1(x^n), \ldots, \phi_k(x^n)$ functions of x^n. Let $A(\lambda) = \ln Z(\lambda)$ and let $\dot{A}(\lambda)$ denote the vector of the derivatives $dA(\lambda)/d\lambda_i$. Then by taking the derivatives with respect to λ in the equation

$$\int f(x^n; \lambda) dx^n = Z(\lambda) \qquad (5.3)$$

we get

$$\dot{A} = \frac{\dot{Z}(\lambda)}{Z(\lambda)} = -E_\lambda \phi(X^n) , \qquad (5.4)$$

which holds for all λ. The maximum-likelihood estimate $\hat{\lambda} = \hat{\lambda}(x^n)$ is seen to satisfy

$$\dot{A}(\hat{\lambda}) + \phi(x^n) = 0 , \qquad (5.5)$$

giving

$$- \ln f(x^n; \hat{\lambda}(x^n)) = A(\hat{\lambda}) + \hat{\lambda}' \phi(x^n) . \qquad (5.6)$$

Next, consider the entropy H_λ

$$-E_f \ln f(X^n; \lambda) = A(\lambda) + \lambda' E_\lambda \phi(X^n) . \qquad (5.7)$$

If we evaluate this at $\hat{\lambda}$ we get with (5.4) and (5.5) the equality

$$H_{\hat{\lambda}} = - \ln f(x^n; \hat{\lambda}) , \qquad (5.8)$$

which shows that the negative logarithm of the maximized-likelihood function has the appearance of the entropy, the so-called empirical entropy, and it depends only on $\hat{\lambda}$; ie, it is the same for all y^n such that $\hat{\lambda}(y^n) = \hat{\lambda}$. Hence, for all such y^n we have the factorization

$$f(y^n; \hat{\lambda}(x^n)) = f(y^n, \hat{\lambda}(x^n)) \tag{5.9}$$

$$= f(y^n | \hat{\lambda}(x^n)) 2^{-h(\hat{\lambda}(x^n))} \tag{5.10}$$

Example 2. The normal density function with $\lambda = 1/(2\sigma^2)$ and $\phi(x^n) = \sum_t (x_t - \mu)^2$ gives $H_{\hat{\lambda}} = (n/2) \ln(2\pi e \hat{\sigma}^2)$.

5.1.2 Maximum entropy family with simple loss functions

An important related class of probability models are induced by loss functions $\delta(y, \hat{y})$; thus

$$-\ln f(y^n | x^n; \lambda, \theta) = n \ln Z(\lambda) + \lambda \sum_t \delta(y_t, \hat{y}_t) , \tag{5.11}$$

where the normalizing coefficient $Z(\lambda)$ does not depend on θ. Such loss functions are called *simple* [22].

We see that if we minimize the ideal code length with respect to λ, we get

$$n d \ln Z/d\lambda + \sum_t \delta(y_t, \hat{y}_t) = 0 \tag{5.12}$$

$$d \ln Z/d\lambda = -\int f(y|x; \lambda, \theta) \delta(y, \hat{y}) dy = -E_{\lambda,\theta} \delta(y, \hat{y}) , \tag{5.13}$$

the latter holding for all λ and θ. As above, the minimizing value $\hat{\lambda}(y^n, x^n) = \hat{\lambda}$ also minimizes the mean $-E_{\lambda,\theta} \ln f(y^n | x^n; \lambda, \theta)$, and for $\lambda = \hat{\lambda}$

$$E_{\lambda,\theta} \sum_t \delta(y_t, \hat{y}_t) = \sum_t \delta(y_t, \hat{y}_t) . \tag{5.14}$$

Hence, having minimized the sum of the prediction errors with respect to both λ and θ, we know this to be the mean performance, the mean taken with the distribution defined by the minimizing parameter value $\hat{\lambda}$ and any θ.

Example 3. As an example consider the quadratic error function. The normalizing integral is given by

$$\int e^{-\lambda(y - F(x,\theta))^2} dy = \sqrt{\pi/\lambda} ,$$

which does not depend on θ. We see that we get a normal distribution with mean $F(x, \theta)$ and variance $\sigma^2 = 1/(2\lambda)$. If we extend this to sequences by independence and minimize the negative logarithm with respect to θ and the variance, we see that the minimizing variance is given by the minimized quadratic loss.

The quadratic error function is but a special case of loss functions of type $\delta(y - \hat{y}) = |y - \hat{y}|^\alpha$ for $\alpha > 0$, which we call the α-class; it is also called the "generalized gaussian" class. The normalizing coefficient is given by

$$\int e^{-\lambda|y-F(x,\theta)|^\alpha} dy = Z_\lambda = \frac{2}{\alpha} \lambda^{-1/\alpha} \Gamma(1/\alpha) , \qquad (5.15)$$

where $\Gamma(x)$ is the gamma-function. Such loss functions are then seen to be simple.

Next we show by generalizing slightly the arguments in [12] that the distributions

$$p(y|x; \lambda, \theta) = Z^{-1}(\lambda) e^{-\lambda\delta(y,\hat{y})} ,$$

where $\hat{y} = h(x, \theta)$, are maximum entropy distributions. Consider the problem

$$\max_g E_g \ln 1/g(Y) , \qquad (5.16)$$

where the maximization is over all g such that $E_g \delta(Y, h(\mathbf{x}; \theta)) \leq E_p \delta (Y, h(\mathbf{x}; \theta))$. We have first by this restriction on the density functions g

$$E_g \ln 1/p(Y|\mathbf{x}; \theta, \lambda) = \lambda E_g \delta(Y, h(\mathbf{x}; \theta)) + \ln Z_\lambda \leq H(p) ,$$

where $H(p)$ denotes the entropy of $p(y|\mathbf{x}; \theta, \lambda)$. Then, by the noiseless coding theorem, the entropy $H(g)$ of g satisfies $H(g) \leq E_g \ln 1/p(Y|\mathbf{x}; \theta, \lambda)$, the right-hand side being upper-bounded by $H(p)$. The equality is reached with $g = p$. This result generalizes the familiar fact that the normal distribution with variance σ^2 has the maximum entropy among all distributions whose variance does not exceed σ^2.

5.2 Universal Models

The fundamental idea with a universal model for a given model class $\mathcal{M}_\gamma = \{f(x^n; \theta, \gamma) : \theta \in \Omega \subset R^k\}$ is that it should assign as large a probability or density to the data strings as we can get with help of the given model class. Equivalently, the ideal code length for the data strings should be minimized. If we are given a particular data string x^n, the universal model must not depend on this string; rather it should depend only on the model class. One way to obtain universal models is to construct a universal coding system by some intuitively appealing coding method, which, in fact, was the way universal coding systems were constructed – and still is to some extend. However, it will be particularly informative and useful to construct optimization problems whose solutions will define universal models.

5.2.1 Mixture models

We construct two universal models as solutions to certain optimization problems, and consider first a mixture universal model. This is perhaps the oldest universal model, also studied in statistics. A prototype of optimization problems for codes is the noiseless coding theorem, expressed as

$$\min_Q \sum_{x^n} P(x^n) \log \frac{P(x^n)}{Q(x^n)} ,$$

which is solved by $Q = P$. Since we are interested in codes for a class of models instead of just one, we may look for distributions $Q(x^n)$ whose ideal code length is the shortest in the mean for the worst case 'opponent' model that generates the data. The first such minmax problem is obtained as follows. Define the "redundancy" as the excess of the mean code length, obtained with a distribution q over the entropy of $f_\theta(x^n) = f(x^n; \theta, \gamma)$, which may be viewed as the conditional density function $f(x^n|\theta, \gamma)$ with the range all the strings of length n,

$$R_n(q, \theta) = \sum_{x^n} f(x^n; \theta, \gamma) \log \frac{f(x^n; \theta, \gamma)}{q(x^n)} = D(f_\theta\|q) . \qquad (5.17)$$

We may then ask for a distribution q as the solution to the minmax problem

$$R_n^+ = \min_q \max_\theta R_n(q, \theta) . \qquad (5.18)$$

This can be tackled more easily by embedding it within a wider problem by considering the mean redundancy

$$R_n(Q, w) = \int w(\theta) R_n(q, \theta) d\theta = \Sigma_{x^n} \int w(\theta) f(x^n|\theta, \gamma) \log \frac{f(x^n|\theta, \gamma)}{q(x^n)} d\theta ,$$

where $w(\theta)$ is a density function for the parameters in a space Ω_γ. There is really no conflict in notations, because the limit of the distributions $w_i(\eta)$ which place more and more of their probability mass in a shrinking neighborhood of θ, call it $w_\theta(\eta)$, will make $R_n(q, w_\theta) = R_n(q, \theta)$. For each w the minimizing distribution of $R_n(q, w)$ is by the noiseless coding theorem the *mixture* distribution

$$f_w(x^n, \gamma) = \int f(x^n; \theta, \gamma) w(\theta) d\theta , \qquad (5.19)$$

which gives the minimized value

$$R_n(w) = \min_q R_n(q, w) = \sum_{x^n} \int w(\theta) f(x^n; \theta, \gamma) \log \frac{f(x^n; \theta, \gamma)}{f_w(x^n, \gamma)} d\theta = I_w(\Theta; X^n) .$$
$$(5.20)$$

This is seen to be the mutual information between the random variables Θ and X^n, defined by the Kullback–Leibler distance between the joint distribution $w(\theta)f(x^n; \theta, \gamma)$ and the product of the marginals $w(\theta)$ and $f_w(x^n)$.

Further, we may ask for the worst case w as follows:

$$\sup_w R_n(w) = \sup_w I_w(\Theta; X^n) = K_n(\gamma) , \qquad (5.21)$$

which is seen to be the capacity of the channel $\Theta \to X^n$. By (11) and (12) in section 1.3 the distance $D(f_\theta \| f_{\bar{w}})$ for the maximizing prior is the same for all θ, which means that $R_n^+ = K_n(\gamma)$ and the model family lies on the surface of a hyperball with the special mixture as the center.

Recently the minmax problem (5.18) has been generalized to the minmax *relative* redundancy problem defined as follows: First, define θ_g as the unique parameter value in Ω^k such that

$$\min_\theta E_g \log \frac{g(X^n)}{f(X^n; \theta, \gamma)} = E_g \log \frac{g(X^n)}{f(X^n; \theta_g, \gamma)} . \qquad (5.22)$$

In words, θ_g picks the model in the class that is nearest to the data-generating model $g(x^n)$ in the Kullback–Leibler distance, which need not lie within the model class $\mathcal{M}_\gamma$. Then consider

$$\min_q \max_g E_g \log \frac{f(x^n; \theta_g, \gamma)}{q(X^n)} . \qquad (5.23)$$

This time the solution is not easy to find, but asymptotically it is a modification of the mixture solving (5.18) [76, 77].

What is the ideal code length of the mixture universal model? It depends on the prior. Asymptotically the maximizing prior $\bar{w}$ (5.21) in the mixture model is Jeffreys' prior,

$$\pi(\hat{\theta}) = \frac{|J(\hat{\theta})|^{1/2}}{\int_\Omega |J(\theta)|^{1/2} d\theta} . \qquad (5.24)$$

The mean of the ideal code length has been shown by several authors for various model classes to be (see, for instance, [10] and [2]),

$$E_\theta \ln \frac{f(X^n; \theta, \gamma)}{f_{\bar{w}}(X^n, \gamma)} = \frac{k}{n} \ln \frac{n}{2\pi} + \ln \int_\Omega |J(\eta)|^{1/2} d\eta + o(1) ,$$

where $J(\theta)$ is the limit of the Fisher information matrix

$$J(\theta) = \lim_{n \to \infty} J_n(\theta) \qquad (5.25)$$

$$J_n(\theta) = -n^{-1} \left\{ E_\theta \frac{\partial^2 \ln f(X^n; \theta, \gamma)}{\partial \theta_i \partial \theta_j} \right\} . \qquad (5.26)$$

The determinant of the Fisher information matrix is called *Fisher information*.

The derivation of the ideal code length of the mixture model is based on Laplace integration, and if we overlook tricky details, the derivation of the ideal code length is easy. Because it is also quite illuminative we outline the derivation. By the Taylor expansion

$$\ln \frac{f(x^n; \hat{\theta}(x^n), \gamma)}{f(x^n; \theta, \gamma)} = \frac{n}{2}(\theta - \hat{\theta}(x^n))' \hat{J}(y^n, \tilde{\theta})(\theta - \hat{\theta}(x^n)) ,$$

where $\hat{J}(y^n, \hat{\theta})$ denotes the empirical Fisher information matrix

$$\hat{J}(y^n, \hat{\theta}) = -n^{-1} \left\{ \frac{\partial^2 \ln f(y^n; \hat{\theta}, \gamma)}{\partial \hat{\theta}_j \partial \hat{\theta}_k} \right\} \tag{5.27}$$

and $\tilde{\theta}$ is between $\hat{\theta}$ and θ. Letting $w(\theta)$ be a smooth prior consider the mixture

$$f_w(x^n, \gamma) \cong f(x^n; \hat{\theta}(x^n), \gamma) \int e^{\frac{-n}{2}(\eta - \hat{\theta}(x^n))' \hat{J}(y^n, \tilde{\theta})(\eta - \hat{\theta}(x^n))} w(\eta) d\eta .$$

For large n, the exponential decreases very rapidly with growing distance $\eta - \hat{\theta}(x^n)$, and if neither the prior nor $\hat{J}(y^n, \hat{\theta})$ changes much in a small neighborhood of $\hat{\theta}(x^n)$, the integral is seen to be approximately

$$\frac{(2\pi)^{k/2}}{n^{k/2} |\hat{J}(\hat{\theta})|^{1/2}} .$$

All told we have

$$\ln 1/f_w(x^n, \gamma) \cong \frac{k}{2} \ln \frac{n}{2\pi} + \ln \frac{|\hat{J}(\hat{\theta})|^{1/2}}{w(\hat{\theta})} .$$

The last term would cause trouble if $\hat{\theta}$ is close to a singularity of the empirical Fisher information matrix, but the danger vanishes if we pick the prior as Jeffreys' prior, (5.24), and we get

$$\ln 1/f_w(x^n, \gamma) \cong \frac{k}{2} \ln \frac{n}{2\pi} + \ln \int_\Omega |J(\eta)|^{1/2} d\eta . \tag{5.28}$$

5.2.2 Normalized maximum-likelihood model

Consider first the minimization problem

$$\min_\theta - \log f(x^n; \theta, \gamma) . \tag{5.29}$$

The minimizing parameter $\hat{\theta} = \hat{\theta}(x^n)$ is the *maximum-likelihood (ML)* esti-mator. Curiously enough, in the previous case the best model is $f(x^n; \theta_g, \gamma)$, but it cannot be computed since g is not known, while here the best code length is obtained with $f(x^n; \hat{\theta}(x^n), \gamma)$, which can be computed, but it is not a model since it does not integrate to unity. However, it may be taken as an ideal target, which suggests the maxmin problem

$$\max_g \min_q E_g \log \frac{f(X^n; \hat{\theta}(X^n), \gamma)}{q(X)} . \tag{5.30}$$

Theorem 16 *The unique solution to this maxmin problem is given by $q(x^n) = g(x^n) = \hat{f}$, where $\hat{f}$ is the normalized maximum-likelihood (NML) universal model*

$$\hat{f}(x^n; \gamma) = \frac{f(x^n; \hat{\theta}(x^n), \gamma)}{C_{\gamma, n}} \tag{5.31}$$

$$C_{\gamma, n} = \int_{\hat{\theta}(y^n) \in \Omega} f(x^n; \hat{\theta}(y^n), \gamma) dy^n \tag{5.32}$$

$$= \int_{\hat{\theta} \in \Omega} g(\hat{\theta}; \hat{\theta}, \gamma) d\hat{\theta} . \tag{5.33}$$

Here

$$g(\hat{\theta}; \theta, \gamma) = \int_{\hat{\theta}(y^n) = \hat{\theta}} f(x^n; \theta, \gamma) dy^n \tag{5.34}$$

denotes the density function of the statistic $\hat{\theta}(y^n)$ induced by the model $f(y^n; \theta, \gamma)$. The maxmin value = minmax value is $\log C_{\gamma, n}$.

For $C_{\gamma, n}$ to be finite the space $\Omega = \Omega_\gamma$ may have to be chosen as an open-bounded set, unless the integral is finite anyway.

Proof. We have

$$\max_g \min_q E_g \log \frac{f(X^n; \hat{\theta}(X^n), \gamma)}{q(X)} = \max_g \min_q \{D(g\|q) - D(g\|\hat{f})\} + \log C_{\gamma, n} . \tag{5.35}$$

Further

$$\max_g \min_q \{D(g\|q) - D(g\|\hat{f})\} \geq \min_q \{D(g\|q) - D(g\|\hat{f})\} \leq 0$$

for all g. The first inequality holds for all g, and the second inequality is true because $D(g\|q)$ is its minimum value zero for $q = g$. The second term $D(g\|\hat{f})$ vanishes for $g = \hat{f}$, and only for this value for g, and the maxmin value (5.35) is $\log C_{\gamma, n}$. If we put $q = \hat{f}$, we see that

$$D(g\|\hat{f}) - D(g\|\hat{f}) = 0$$

for all g, and since minmax is not smaller than maxmin even the minmax value is $\log C_{\gamma, n}$ as claimed. We consider the maxmin problem more important because of the uniqueness of the solution.

Both of the maxmin and minmax problems are related to Shtarkov's min-max problem [69]:

$$\min_{q} \max_{x^n} \log \frac{f(x^n; \hat{\theta}(x^n), \gamma)}{q(x^n)}$$

with the unique solution $\hat{q} = \hat{f}$. Indeed, the ratio $f(x^n; \hat{\theta}(x^n), \gamma)/C_{\gamma,n}q(x^n)$ of two density functions must exceed unity at some point, unless the two are identical, and the maximum ratio reaches its minimum, unity, for the *NML* density function.

We outline the derivation of a quite accurate formula for the negative logarithm of the NML model. The main condition required is that the ML-estimates satisfy the central limit theorem (CLT) in that $\sqrt{n}(\hat{\theta}(x^n) - \theta)$ converges in distribution to the normal distribution of mean zero and covariance $J^{-1}(\theta)$, where

$$J(\theta) = \lim_{n \to \infty} J_n(\theta) \tag{5.36}$$

$$J_n(\theta) = -n^{-1} \left\{ E_\theta \frac{\partial^2 \ln f(X^n; \theta, \gamma)}{\partial \theta_i \partial \theta_j} \right\} . \tag{5.37}$$

$J_n(\theta$ is the Fisher information matrix (5.26). The derivation is particularly simple if we assume the convergence

$$g(\hat{\theta}; \hat{\theta})(\frac{n}{2\pi})^{-k/2} \to |J(\hat{\theta})|^{1/2} , \tag{5.38}$$

which does not follow from the CLT in the form stated. The proof without this assumption is more complex; it is given in [60]. Also,

$$\pi_n(\hat{\theta}) = \frac{g(\hat{\theta}; \hat{\theta})}{C_n} \to \pi(\hat{\theta}) = \frac{|J(\hat{\theta})|^{1/2}}{\int_\Omega |J(\theta)|^{1/2} d\theta} . \tag{5.39}$$

We call $\pi_n(\theta)$ the *canonical prior;* the limit is Jeffreys' prior (5.24).

We have then the desired asymptotic formulas

$$-\log \hat{f}(x^n; \mathcal{M}_\gamma) = -\log f(x^n; \hat{\theta}(x^n), \gamma) + \frac{k}{2} \log \frac{n}{2\pi} + \log \int_\Omega |J(\theta)|^{1/2} d\theta + o(1) . \tag{5.40}$$

The term $o(1)$, which goes to zero as $n \to \infty$, takes care of the rate of convergence by the CLT and other details.

It is important that the optimal $q = \hat{f}$ be definable in terms of the model class $\mathcal{M}_\gamma$ selected, and hence be computable, although the computation of the integral of the square root of the Fisher information may pose a problem. There are a number of ways to approximate it; an important one is presented in [46].

The minmax problem (5.23) and the maxmin problem (5.30) are related and their solutions should be close, at least asymptotically. This can be illustrated by the following theorem, which in essence is due to P. Grünwald and H. White [22, 84].

Theorem 17 *Let the data be generated by an iid process $g(\cdot)$. Then with probability unity under g*

$$\hat{\theta}(x^n) \to \theta_g \ . \tag{5.41}$$

Further, if $\{q_n(x^n)\}$ is a family of density functions, then for every n,

$$\frac{1}{m} \sum_{t=0}^{m-1} \ln \frac{f(x_{tn+1}^{(t+1)n}; \hat{\theta}(nm), \gamma)}{q_n(x_{tn+1}^{(t+1)n})} \to E_g \ln \frac{f(X^n; \theta_g, \gamma)}{q_n(X^n)} \tag{5.42}$$

in g-probability 1, as $m \to \infty$, where $\hat{\theta}(nm)$ denotes the ML estimate evaluated from x_1^{nm}.

Sketch of Proof. We have first

$$E_g \frac{\partial}{\partial \theta} \ln f(X; \theta_g, \gamma) = 0 \ , \tag{5.43}$$

where the argument θ_g indicates the point where the derivative is evaluated. Similarly,

$$\frac{1}{n} \sum_{t=1}^{n} \frac{\partial}{\partial \theta} \ln f(x_t; \hat{\theta}(n), \gamma) = 0 \ .$$

For each fixed value $\hat{\theta}(n) = \theta$ this sum converges by the strong law of large numbers to the mean $E_g \log f(X; \theta)$, in g-probability 1, and the sum is always zero, which in light of (7.4) means that the limiting mean is zero. This proves (5.41). We can again apply the strong law of large numbers to (5.42), and in view of (5.41) conclude the second claim.

How about the *NML* distribution for the class $\mathcal{M} = \bigcup_\gamma \mathcal{M}_\gamma$? It turns out that although we cannot meaningfully maximize $f(x^n; \hat{\theta}(x^n), \gamma)$ over γ in its range Γ we can maximize $\hat{f}(x^n; \mathcal{M}_\gamma)$, which leads to the *NML* distribution

$$\hat{f}(x^n; \Gamma) = \frac{\hat{f}(x^n; \hat{\gamma}(x^n))}{\sum_{y^n : \hat{\gamma}(y^n) \in \Gamma} \hat{f}(y^n; \hat{\gamma}(y^n))} \ . \tag{5.44}$$

We can write the denominator in the following form

$$\sum_{y^n} \hat{f}(y^n; \hat{\gamma}(y^n)) = \sum_\gamma Q(\gamma) \tag{5.45}$$

$$Q(\gamma) = \sum_{\hat{\gamma}(y^n) = \gamma} \hat{f}(y^n; \hat{\gamma}(y^n)) \tag{5.46}$$

and

$$\hat{f}(x^n; \Gamma) = \frac{\hat{f}(x^n; \hat{\gamma}(x^n))}{Q_n(\hat{\gamma}(x^n))} q_n(\hat{\gamma}(x^n)) \ , \tag{5.47}$$

where

$$q_n(\gamma) = \frac{Q_n(\gamma)}{\sum_{\gamma' \in \Gamma} Q_n(\gamma')} \tag{5.48}$$

is seen to sum up to unity.

Is there anything we can say about the distribution $q_n(\gamma)$? In fact, we have the following theorem:

Theorem 18 *Let for each* $\gamma \in \Gamma_f$, $Prob(\hat{\gamma}(y^n) \neq \gamma) \to 0$, *where* Γ_f *is a finite subset of* Γ *and* $\theta \in \Omega_\gamma$. *Then*

$$Q_n(\gamma) \to 1 \tag{5.49}$$
$$q_n(\gamma) \to 1/|\Gamma_f| . \tag{5.50}$$

Proof. We have

$$1 = \sum_{y^n} \hat{f}(y^n; \gamma) = Q_n(\gamma) + \sum_{\hat{\gamma}(y^n) \neq \gamma} \hat{f}(y^n; \hat{\gamma}(y^n)) , \tag{5.51}$$

where the second sum converges to zero by assumption. The claim follows.

5.2.3 A predictive universal model

Although the *NML* universal model appears to capture properties of data in a superb manner its application has two problems: First, the normalizing coefficient can be evaluated only for restricted classes of models. Secondly, it does not define a random process, and hence it cannot be applied to prediction.

In this subsection we describe a predictive universal model, called a pre-quential model in [15], and predictive MDL model in [56], which requires no normalization. Consider the conditional probabilities or densities

$$f(x_{t+1}|x^t, \bar{\theta}(x^t), \gamma) = \frac{f(x^{t+1}; \bar{\theta}(x^t), \gamma)}{\int f(x^t, u; \bar{\theta}(x^t), \gamma) du} ,$$

where $\bar{\theta}(x^t)$ is an estimate of θ. It will be taken close to the ML estimate $\hat{\theta}(x^t)$, but since for small values of t we may not be able to obtain the ML estimate with a selected number of components, we need another estimate with fewer components. Define the universal model $\bar{f}(x^n)$ for the class $\mathcal{M}_\gamma$ by its negative logarithm as follows:

$$- \log \bar{f}(x^n; \gamma) = - \sum_t \log f(x_{t+1}|x^t, \bar{\theta}(x^t), \gamma) .$$

By summation by parts

$$- \log \bar{f}(x^n; \gamma) = - \log f(x^n; \bar{\theta}(x^n), \gamma) + \sum_{t=0}^{n-1} \log \frac{f(x^{t+1}; \bar{\theta}(x^{t+1}, \gamma))}{f(x^{t+1}; \bar{\theta}(x^t), \gamma)}. \tag{5.52}$$

Since the ratios in the sum exceed unity, except when $\bar{\theta}(x^{t+1})$ happens to equal $\bar{\theta}(x^t)$, the sum is positive. It represents the code length for the model $\bar{\theta}(x^n)$ although in an implicit manner since no actual coding of the parameters are done. They are determined from the past by an algorithm which the decoder is supposed to know.

We prove in the next section a generalization of the noiseless coding theorem, which implies

$$E_{\theta,\gamma} \sum_{t=0}^{n-1} \log \frac{f(x^{t+1}; \hat{\theta}(x^{t+1}); \gamma)}{f(x^{t+1}; \hat{\theta}(x^t); \gamma)} \geq \frac{k - \epsilon}{2} \log n ,$$

for all positive numbers ϵ, n large enough, and all θ except in a set whose volume goes to zero as n grows. Here, the expectation is with respect to the model $f(x^n; \theta, \gamma)$. The predictive universal code has been shown to reach the bound in the right hand side for linear quadratic regression problems of various kinds, see [57, 14, 82]. In a later subsection we construct a predictive universal model for the Tree Machines, which also turns out to be asymptotically optimal.

Typically a predictive universal model is appropriate for time series, for it requires an ordering of the data, which, of course can be done for any data, although the result depends somewhat on the way the data are ordered. A reasonable rule for ordering is to give priority to data that are easy to predict or encode. Despite the initialization problem, which is less important for large amounts of data, the predictive universal model has the advantage that it does not require the Fisher information matrix nor the integral of the square root of its determinant not other details of how the models are encoded. This turns out to be important for the Markov chains and the tree machines, for which the integral is difficult to carry out.

With the further generalization of replacing the fixed γ in (5.52) by $\bar{\gamma}(x^t)$ we get a powerful universal model

$$\begin{aligned} -\log \bar{f}(x^n; \bar{\gamma}(x^n)) &= -\log f(x^n; \bar{\theta}(x^n), \bar{\gamma}(x^n)) \\ &+ \sum_{t=0}^{n-1} \log \frac{f(x^{t+1}; \bar{\theta}(x^{t+1}, \bar{\gamma}(x^{t+1}))}{f(x^{t+1}; \bar{\theta}(x^t), \bar{\gamma}(x^t))} \end{aligned} \tag{5.53}$$

for the entire model class $\mathcal{M}$. Depending on how we pick the structure indices $\bar{\gamma}(x^t)$ we can capture subtle additional regular features in the data which neither one of the two other universal models can do. We discuss an example in the chapter the chapter 8, The MDL Principle.

There has been recent improvement of the predictive principle based on the following idea: Since learning good parameter values requires a lot of data a model with a suitably selected fixed parameter value can predict better up to a certain point, after which the recursively improving parameter values should be switched on; for details we refer to [19].

Table 5.1. Weights of 13 mice at 7 time points

Days after birth						
3	6	9	12	15	18	21
0.109	0.388	0.621	0.823	1.078	1.132	1.191
0.218	0.393	0.568	0.729	0.839	0.852	1.004
0.211	0.394	0.549	0.700	0.783	0.870	0.925
0.209	0.419	0.645	0.850	1.001	1.026	1.069
0.193	0.362	0.520	0.530	0.641	0.640	0.751
0.201	0.361	0.502	0.530	0.657	0.762	0.888
0.202	0.370	0.498	0.650	0.795	0.858	0.910
0.190	0.350	0.510	0.666	0.819	0.879	0.929
0.219	0.399	0.578	0.699	0.709	0.822	0.953
0.255	0.400	0.545	0.690	0.796	0.825	0.836
0.224	0.381	0.577	0.756	0.869	0.929	0.999
0.187	0.329	0.441	0.525	0.589	0.621	0.796
0.278	0.471	0.606	0.770	0.888	1.001	1.105

Example 1

We illustrate the predictive principle by a simple example. Table 5.1 [48] consists of the weights of 13 mice, obtained in seven consecutive time instances at 3-day intervals since birth. The objective is to predict the weights in the last column from the observations in the first six using polynomial models. The prediction error is measured by the sum of the squares of the deviations for all the mice, while the polynomials may be fitted using any criterion.

We consider first the model class obtained when a k-degree polynomial is fitted to each row, to the weights of each mouse. The weight y_{it} of the mouse i at tth weighing instance is given by

$$y_{it} = \sum_{j=0}^{k} \theta_j t^j + \epsilon_{it}, \quad i = 1, \ldots, 13, \ t = 1, \ldots, 7, \tag{5.54}$$

where ϵ_{it} denotes a family of zero-mean uncorrelated random variables with variance σ^2, also a parameter. The parameters to be estimated are k, σ, and $\theta_0, \ldots, \theta_k$. The estimate of the degree k will be common for all the mice, but the coefficients are estimated for each mouse individually by the least-squares technique.

The estimation of the optimal degree is an insecure matter by traditional means, and in fact we get different values using different techniques. However, with the hindsight offered by a peek at the weights in the last column to be predicted, Rao verified that a quadratic polynomial will yield the best prediction. For the first row, for example, the best least-squares polynomial fitted to the first six measurements is $f_1(t) = -0.219 + 0.341t - 0.0185t^2$. Hence the prediction of the last weight is then $f_1(7) = 1.2615$ with the error $\hat{e}_{17} = 0.033$.

We wish to determine the optimal degree by an application of the predictive *MDL* principle to the first six columns of Table 5.1. By (5.54) we have first the code length for the ith row,

$$L(y_i|k) = \sum_{t=1}^{6} \hat{e}_{it}^2 \,,$$

where $\hat{e}_{it} = y_{it} - \sum_{j=0}^{k} t^j \theta_j(t-1)$, and $\hat{\theta}_j(t-1), j = 1, \ldots, k \wedge t - 1$ denote the least-squares estimates obtained from the first $t - 1$ columns; we denote by $a \wedge b$ the minimum of the two numbers. The very first column is predicted as zero. The criterion to minimize for k is

$$L(k) = \sum_{i=1}^{13} L(y_i|k) \,.$$

The numerical values are $L(0) = 0.498, L(1) = 0.101, L(2) = 0.095, L(3) = 0.122, L(4) = 0.222$, which give the optimal degree $k = 2$ in agreement with the optimum determined with the aid of hindsight.

With the optimal degree thus established, sophisticated techniques developed in traditional statistics can now be put to work in order to find the best way to estimate the coefficients of the quadratic polynomial, which by no means needs to be the least-squares fit. In Rao [48] five methods and the resulting predictors for the weights in the last column of Table 5.1 were worked out – namely, BLUP (best linear unbiased predictor), JSRP (James–Stein regression predictor) [32], SRP (shrunken regression predictor), EBP (empirical bayes predictor) [47], and RRP (ridge regression predictor) [31] – which gave the average square prediction errors shown in Table 5.2.

Table 5.2. Prediction errors with various predictors

$BLUB$	JSRP	SRP	EBP	RRP
0.1042	0.1044	0.0970	0.0950	0.1050

We see that the difference between the best and the worst is about 5 percent, which means that we are talking about marginal effects.

Quite interestingly, Rao also noticed that one could fit the polynomials only to the past q columns, where q should be determined optimally. When he did this in hindsight with the knowledge of the weights in the seventh column to be predicted, the best predictions were found to result from the line that for each mouse – i.e., for each row in Table 5.1 – passes through the fifth and the sixth weights. Hence, for example, for the first row the best prediction is $1.132 + (1.132 - 1.078) = 1.186$. This time the average squared prediction error was much better, namely, 0.0567. Hence, we see that the initial traditional assumption of the parent population and distribution was not too good after all, not even when the increasingly sophisticated refinements developed in the field have been invoked. In our language, what Rao did was to consider different model families M_q, one for each q.

Since we plan to apply the predictive MDL principle, we have only the first six columns available, which means that q ranges from 1 to 5. Indeed, the first family examined above is M_5, within which we found the optimal class to be obtained for $k = 2$. The criterion, written as $L(k, q)$, is the same as used above, except that the least-squares estimates $\hat{\theta}_j(t - 1|k, q), j = 1, \ldots, \min\{k, q, t - 1\}$ are calculated from the columns $t - q, \ldots, t - 1$, whenever possible. The numerical values for $q \geq 2$ are as follows: $L(0, 1) = 0.150$, $L(0, 2) = 0.253$, $L(1, 2) = 0.0944$, $L(0, 3) = 0.359$, $L(1, 3) = 0.0944$, $L(2, 3) = 0.108$, $L(0, 4) = 0.448$, $L(1, 4) = 0.0977$, $L(2, 4) = 0.0963$, $L(3, 4) = 0.148$, while the values of $L(k, 5) = L(k)$ were given above for the first model family. We see that the overall best model classes are obtained for $k = 1$ with q either 2 or 3, so that the linear predictor is the best, and it is to be fitted either to the past two or three weights. Except for the tie, both of these optimal degrees of polynomials in the two considered model classes agree with those found in Rao [48], with the help of hindsight.

When the two line-predictors, both optimal in light of the available six columns, were applied to predicting the values in the seventh column, the average squared prediction errors were 0.0567, obtained with the lines passing through the past two points, and 0.0793, obtained with the lines fitted to the past three points for each mouse. Both values are well below those obtained with the quadratic polynomial, fitted to all the past data, regardless of how the values of the parameters are calculated. We then conclude that hindsight was not needed, and that the predictive MDL principle did give us a clear indication of how the weights should be predicted. We do not know, however, whether even better predictive models could be found by using ones imagination, but fine-tuning the parameter estimates appears to be of secondary importance.

5.2.4 Conditional NML model

We conclude the discussion of universal models with an outline of a recent discovery, which at this writing is still under development. Consider the model class $\{f(x^n; \theta)\}$, $\theta = \theta_1, \ldots, \theta_k$, and the ML estimate $\hat{\theta}(x^n)$. Regard the maximized likelihood $f(x^{n-1}, x_n; \hat{\theta}(x^{n-1}, x_n))$ as a function of the last symbol x_n only. Construct the conditional density function

$$f(x_t|x^{t-1}) = \frac{f(x^{t-1}, x_t; \hat{\theta}(x^{t-1}, x_t))}{K_t} \tag{5.55}$$

$$K_t = \int f(x^{t-1}, x; \hat{\theta}(x^{t-1}, x))dx. \tag{5.56}$$

This defines the density function for the data

$$f(x^n) = \prod_{t_k}^{n} f(x_t|x^{t-1})p(x^{t_k}) \qquad (5.57)$$

where t_k is the smallest value of t for which the ML estimate is defined and $p(x^{t_k})$ a suitably calculated initial density.

The *conditional NML* model $f(x_t|x^{t-1})$ is the unique solution to the minmax problem

$$\min_{q(x|x^{t-1})} \max_{x,\theta} \log \frac{f(x^{t-1}, x; \theta)}{q(x|x^{t-1})}, \qquad (5.58)$$

which is proved the same way as Shtarkov's minmax problem above.

Example. For the Bernoulli class with $\hat{\theta}(x^n) = n_1/n$, where $n_1 = \sum_t x_t$ is the number of 1's in x^n. If $n_0 = n - n_1$ the maximized likelihood is

$$P(x^n; n_1/n) = \left(\frac{n_1}{n}\right)^{n_1} \left(\frac{n_0}{n}\right)^{n_0}.$$

The conditional *NML* probability is given by

$$P(1|x^n) = \frac{\left(\frac{n_1+1}{n+1}\right)^{n_1+1} \left(\frac{n_0}{n+1}\right)^{n_0}}{\left(\frac{n_1+1}{n+1}\right)^{n_1+1} \left(\frac{n_0}{n+1}\right)^{n_0} + \left(\frac{n_0+1}{n+1}\right)^{n_0+1} \left(\frac{n_1}{n+1}\right)^{n_1}}$$

$$= \frac{(n_1+1)(1+1/n_1)^{n_1}}{(n_0+1)(1+1/n_0)^{n_0} + (n_1+1)(1+1/n_1)^{n_1}}. \qquad (5.59)$$

We take here $(1 + 1/k)^k = 1$ for $k = 0$. Notice that the probability of the first symbol, too, is the solution to the minmax problem: since the maximized likelihood for the first symbol is 1 we get $P(x_1) = 1/(1+1) = 1/2$.

For short strings or strings where the ratio n_1/n is close to 0 or 1, the conditional probability 5.59 behaves like the Krichevski-Trofimov predictive probability

$$P_{KT}(1|x^n) = \frac{n_1 + 1/2}{n+1},$$

and for other strings since $(1 + 1/k)^k \to e$ as k grows, it behaves like the Laplace predictive probability

$$P_L(1|x^n) = \frac{n_1 + 1}{n + 2},$$

neither of which has any particular optimality property.

In [78] the same conditional probability 5.59 was obtained as a solution to the minmax problem

$$\min_{\theta} \max_{x} \log \frac{f(x^{t-1}, x; \hat{\theta}(x^n))}{f(x, x^{t-1}, \theta)}. \qquad (5.60)$$

This in general is a difficult problem requiring restrictions on the range of the parameters, and the rather complex solution obtained by the authors was done by regarding the Bernoulli class as a member of the exponential family. That the two solutions agree for the Bernoulli class is because the solution 5.59 to the wider minmax problem 5.58 happens to belong to the Bernoulli class. In general, of course, the solutions to the two minmax problems are different.

5.3 Strong Optimality

The two universal models considered as solutions to the minmax and maxmin problems, while certainly optimal for the worst-case opponent, leave open the nagging question whether their ideal code length would be shorter for all the other opponents or perhaps for most of them. We give a fundamental theorem for universal models [55], which may be viewed as an extension of the first part of the noiseless coding theorem, and it shows that the worst-case performance is the rule rather than an exception. For this we consider a class of parametric probability density functions $\mathcal{M}_\gamma = \{f(x^n; \theta, \gamma)\}$, where the parameter vector $\theta = \theta_1, \ldots, \theta_k$ ranges over an open-bounded subset $\Omega = \Omega^k$ of the k-dimensional euclidian space. The parameters are taken to be "free" in the sense that distinct values for θ specify distinct probability measures.

Theorem 19 *Assume that there exist estimates $\hat{\theta}(x^n)$ which satisfy the central limit theorem at each interior point of Ω^k, where $k = k(\gamma)$, such that $\sqrt{n}(\hat{\theta}(x^n) - \theta)$ converges in probability to a normally distributed random variable. If $q(x^n)$ is any density function defined on the observations, then for all positive numbers ϵ and all $\theta \in \Omega^k$, except in a set whose volume goes to zero as $n \to \infty$,*

$$E_\theta \log \frac{f(x^n; \theta, \gamma)}{q(x^n)} \geq \frac{k - \epsilon}{2} \log n.$$

The mean is taken relative to $f(x^n; \theta, \gamma)$.

We give the original proof, which can be generalized to the proof of a stronger version of the theorem [56]. There exists an elegant proof of an extension of the theorem in [42].

Proof. Consider a partition of the set Ω^k into k-dimensional hypercubes of edge length $\Delta_n = c/\sqrt{n}$, where c is a constant. Let the, say, m_n, centers of these hypercubes form the set $\Omega(\Delta_n) = \{\theta^1, \theta^2, \ldots\}$ and write $C_n(\theta)$ for the cube with center at θ, $\theta \in \Omega(\Delta_n)$. We need to construct a corresponding partition of a subset of X_n, the set of all sequences of length n. For this we use the estimator, $\hat{\theta}(x^n)$, and we define $X_n(\theta) = \{x^n | \hat{\theta}(x^n) \in C_n(\theta)\}$. Let the probability of this set under the distribution $f(x^n; \theta, \gamma)$ be $P_n(\theta)$. Because of the assumed consistency of the estimator, the probability of the set $X_n(\theta)$, namely $P_n(\theta)$, satisfies the inequality

$$P_n(\theta) \geq 1 - \delta(c) \tag{5.61}$$

for all n greater than some number, say, n_c, that depends on c determining the size of the equivalence classes in the partition. Moreover, $\delta(c)$ can be made as small as we please by selecting c large enough.

Consider next a density function $q(x^n)$, and let $Q_n(\theta)$ denote the probability mass it assigns to the set $X_n(\theta)$. The ratio $f(x^n; \theta, \gamma)/P_n(\theta)$ defines a distribution on $X_n(\theta)$, as does of course $q(x^n)/Q_n(\theta)$. By the noiseless coding theorem, applied to these two distributions, we get

$$\int_{X_n(\theta)} f(x^n; \theta, \gamma) \log \frac{f(x^n; \theta, \gamma)}{q(x^n)} dx^n \geq P_n(\theta) \log \frac{P_n(\theta)}{Q_n(\theta)} . \tag{5.62}$$

For a positive number ϵ, $\epsilon < 1$, let $A_\epsilon(n)$ be the set of θ such that the left-hand side of (5.62), denoted $T_n(\theta)$, satisfies the inequality

$$T_n(\theta) < (1 - \epsilon) \log n^{k/2} . \tag{5.63}$$

From (5.62) and (5.63) we get

$$- \log Q_n(\theta) < [(1 - \epsilon)/P_n(\theta) - \frac{\log P_n(\theta)}{\log n^{k/2}}] \log n^{k/2} , \tag{5.64}$$

which holds for $\theta \in A_\epsilon(n)$. Replace $P_n(\theta)$ by its lower bound $1 - \delta(c)$ in (5.61), which does not reduce the right-hand side. Pick c so large that $\delta(c) \leq \epsilon/2$ for all n greater than some number n_ϵ. The first term within the bracket is then strictly less than unity, and since the second term is bounded from above by $(- \log(1 - \epsilon/2))/\log n^{k/2}$, and hence converges to zero with growing n, the expression within the brackets is less than some α such that $0 < \alpha < 1$, for all sufficiently large n, say, larger than some n'_ϵ. Therefore,

$$Q_n(\theta) > n^{-\alpha k/2} \tag{5.65}$$

for $\theta \in A_\epsilon(n)$ and n larger than n'_ϵ. Next, let $B_\epsilon(n)$ be the smallest set of the centers of the hypercubes which cover $A_\epsilon(n)$, and let ν_n be the number of the elements in $B_\epsilon(n)$. Then the Lebesgue volume V_n of $A_\epsilon(n)$ is bounded by the total volume of the ν_n hypercubes, or

$$V_n \leq \nu_n c^k / n^{k/2} . \tag{5.66}$$

From (5.65) and the fact that the sets $X_n(\theta)$ are disjoint we get further

$$1 \geq \sum_{\theta \in B_\epsilon(n)} Q_n(\theta) \geq \nu_n n^{-\alpha k/2} , \tag{5.67}$$

which gives an upper bound for ν_n. From (5.66) we then get the desired inequality

$$V_n \le c^k n^{(\alpha-1)k/2} , \tag{5.68}$$

which shows that $V_n \to 0$ as n grows to infinity.

Using the inequality $\ln z \ge 1 - 1/z$ for $z = f(x^n; \theta, \gamma)/q(x)$ we get

$$\int_{\bar{X}_n(\theta)} f(x^n; \theta, \gamma) \ln[f(x^n; \theta, \gamma)/q(x^n)]dx^n \ge Q_n(\theta) - P_n(\theta) > -1 , \tag{5.69}$$

where $\bar{X}$ denotes the complement of X. To finish the proof, let $\theta \in \Omega^k - A_\epsilon(n)$. Then the opposite inequality, $\ge$, in (5.63) holds. By adding the left-hand sides of (5.62) and (5.69) we get

$$E_\theta \ln[f(x^n; \theta, \gamma)/q(x^n)] > (1 - \epsilon) \ln n^{k/2} - 1,$$

which concludes the proof.

We add that with further and not too severe smoothness conditions on the models one can also prove with the Borel–Cantelli lemma a similar theorem in the almost sure sense: Let $Q(x^n)$ define a random process. Then for all $\theta \in \Omega^k$, except in a set of measure 0,

$$-\log Q(x^n) \ge -\log P(x^n; \theta, \gamma) + \frac{k - \epsilon}{2} \log n \tag{5.70}$$

for all but finitely many n in $P(x^\infty; \theta)$ – probability 1. Here, $P(x^\infty; \theta)$ denotes the unique measure on infinite sequences defined by the family $\{P(x^n; \theta, \gamma)\}$.

These results show that the right-hand side bound cannot be beaten by any code, except one in the necessary loophole of Lebesgue measure zero, while reachability results immediately from (5.40). Clearly, such an exception must be provided, because the data-generating model itself gives a shorter code length, but it, of course, could not be found except with a wild guess. Moreover, since the parameter value of the data-generating model would have to be communicated to the decoder, it would have to be computable. Since there are at most a countable number of computable numbers, we see that, indeed, the inequality in the theorem is virtually impossible to beat! There exists also a stronger version of the theorem about the minmax bound $\log C_{k,n}$, which in effect states that this worst case bound is not a rare event but that it cannot be beaten in essence even when we assume that the data were generated by the most "benevolent" opponents [62].

5.3.1 Prediction bound for α-loss functions

Consider the α-loss function $\delta(y - \hat{y}) = |y - \hat{y}|^\alpha$ for $\alpha > 0$, where $\hat{y}_t = f_t(y^{t-1}, \mathbf{x}^n)$, $f_1(y^0, \mathbf{x}^n) = 0$, is any predictor. The normalizing coefficient is given by (5.15)

$$\int e^{-\lambda|y-\mu|^\alpha} dy = Z_\lambda = \frac{2}{\alpha} \lambda^{-1/\alpha} \Gamma(1/\alpha) . \tag{5.71}$$

For $\mu = \theta'\mathbf{x}_t$ we have the induced parametric models

$$p(y^n|\mathbf{x}^n;\lambda,\theta) = Z^{-n}(\lambda)e^{-\lambda L(Y^n|\mathbf{x}^n;\theta)} ,$$

where

$$L(Y^n|\mathbf{x}^n;\theta) = \sum_t \delta(y_t,\theta'\mathbf{x}_t)$$

and which we write p_θ for short. Let $\mathcal{M}_{\lambda,\alpha}$ denote the class of such models for $\theta \in \Omega \subset R^k$.

For the predictor $\mu = f_t(y^{t-1},\mathbf{x}^n)$ we put

$$L_f(y^n|\mathbf{x}^n) = \sum_t \delta(y_t,\hat{y}_t) , \tag{5.72}$$

and we get the induced model p_f.

Consider the minmax problem

$$\min_f \max_{p_\theta} E_{p_\theta}[L_f(Y^n|\mathbf{x}^n) - L(Y^n|\mathbf{x}^n;\hat{\theta}(Y^n,\mathbf{x}^n))] . \tag{5.73}$$

Theorem 20 *Let $\hat{y}_t = f_t(y^{t-1},\mathbf{x}^n)$ be any predictor. Then for all α in the interval $[1,2]$ and all positive ϵ, the inequality*

$$\frac{1}{n}E_{p_\theta}L_f(Y^n|\mathbf{x}^n) \geq \frac{1}{\alpha\lambda}(1 + (k-\epsilon)\frac{\alpha}{2n}\ln n) \tag{5.74}$$

holds for n large enough and for all $\theta \in \Omega$, except in a set whose volume goes to zero as n grows to infinity; the expectation is with respect to p_θ.

Proof. Consider

$$E_{p_\theta}\ln\frac{p_\theta}{p_f} = \lambda E_{p_\theta}(L_f(Y^n|\mathbf{x}^n) - L(Y^n|\mathbf{x}^n;\theta)) .$$

For $\alpha \in [1,2]$ one can show that the ML estimates satisfy the central limit theorem and the conditions in the previous theorem are satisfied. The right-hand side exceeds $\frac{k-\epsilon}{2}\ln n$ with the quantifications given. As in (5.13) we have

$$E_{p_\theta}L(Y^n|\mathbf{x}^n;\theta) = \frac{n}{\alpha\lambda} ,$$

and the claim follows.

In light of the maxmin Theorem 14 and the strong optimality theorem in the previous subsection we are justified to regard $-\log\hat{f}(x^n;\gamma)$ (5.40),

$$-\log\hat{f}(x^n;\gamma) = -\log f(x^n;\hat{\theta}(x^n),\gamma) + \log C_{\gamma,n} \tag{5.75}$$

$$\log C_{\gamma,n} = \frac{k}{2}\log\frac{n}{2\pi} + \log\int_\Omega |I(\theta)|^{1/2}d\theta + o(1) \tag{5.76}$$

as the *stochastic complexity* of x^n, given the model class $\mathcal{M}_\gamma$.

Historical Notes

The mixture universal model has been known for a long time, both in information theory and Bayesian statistics, although the mean ideal code length has been derived only relatively recently; see, for instance, [10] and [2] to mention just two.

The predictive principle was invented at about the same time, the early 1980s, by P. Dawid as the *prequential* principle [15] and myself simply as the predictive MDL principle. Although I did not publish the idea, there is a clear reference to it in [53]. I was led to the idea by trying to model gray-level images by passing a plane through the three past nearest pixels to the current one with which to predict it. The plane did a fair job but I thought that by fitting a second-degree function would do a better one. To my surprise the second degree function did a poorer job than the plane, until I understood that the reason was that I had fitted both models to past data. Curiously enough the predictive least-squares technique was applied in the control theory literature at the time to facilitate the calculations of the ML estimates of the parameters in ARMA models. No one seemed to realize that the same technique could be applied to estimate the number of the parameters.

The *NML* universal model was introduced as a universal code apparently by Shtarkov [69]. Its use as a universal model was introduced and its ideal code length derived in [60]. The name *NML model* was invented in [7].

6

Structure Function

We extend Kolmogorov's structure function in the algorithmic theory of complexity to statistical models in order to avoid the problem of noncomputability. For this we have to construct the analog of Kolmogorov complexity and to generalize Kolmogorov's model as a finite set to a statistical model. The Kolmogorov complexity $K(x^n)$ will be replaced by the *stochastic complexity* for the model class $\mathcal{M}_\gamma$ (5.40) and the other analogs required will be discussed next. In this section the structure index γ will be held constant, and to simplify the notations we drop it from the models written now as $f(x^n; \theta)$, and the class as $\mathcal{M}_k$. The parameters θ range over a bounded subset Ω of R^k.

6.1 Partition of Parameter Space

The restrictions imposed by estimability and coding of the parameters require us to consider finite subsets of models, which means that the parameters must be quantized. We also need to measure the distance between two models, which can be done either in the parameter space or in the set of the models themselves. The appropriate distance measure for the models is the KL distance

$$D(f(X^n; \theta) \| f(X^n; \theta')) = E_\theta \log \frac{f(X^n; \theta)}{f(X^n; \theta')} \ .$$

By Taylor series expansion we have

$$\frac{1}{n} D((f(X^n; \theta) \| f(X^n; \theta')) = (\theta' - \theta)^T J_n(\theta)(\theta' - \theta) + O((\theta' - \theta)^3) \ ,$$

where $J_n(\theta)$ is the Fisher information matrix

$$J_n(\theta) = \frac{1}{n} E_\theta \left\{ \frac{\partial^2 \ln 1/f(X^n; \theta)}{\partial \theta_j \partial \theta_k} \right\} \tag{6.1}$$

assumed to exist for all $\theta \in \Omega$. The expectation is with respect to the distribution $f(x^n; \theta)$. We then see that the corresponding distance measure in the parameter space is the one induced by the Fisher information matrix, or its limit $\lim J_n(\theta) = J(\theta)$ as $n \to \infty$. It is clear that the parameter space will have to be such that it includes no singular points of the information matrix.

To a partition of the model space $\mathcal{M}_k$ there corresponds a partition of the parameter space, and picking a representative θ^i of an equivalence class in the latter gives one, $f(x^n; \theta^i)$, in the former. We want a partition in the model space such that the KL distance between any two adjacent models is constant. To achieve this approximately for finite values of n and increasingly accurately as n grows, consider a hyper-ellipsoid $D_{d/n}(\theta^i)$ in the bounded parameter space, centered at θ^i, and defined by

$$(\theta - \theta^i)' J(\theta^i)(\theta - \theta^i) = d/n , \tag{6.2}$$

where d is a parameter. We now indicate the transpose by a prime.

Let $B_{d/n}(\theta^i)$ be the largest rectangle within $D_{d/n}(\theta^i)$ centered at θ^i (see Figure 6.1 below). Its volume is

$$|B_{d/n}(\theta^i)| = \left(\frac{4d}{nk}\right)^{k/2} |J(\theta^i)|^{-1/2} = 2^k \prod_{j=1}^{k} \mu_j , \tag{6.3}$$

where $\mu_j = \sqrt{d}/\sqrt{nk\lambda_j}$ is one-half of the j'-th side length of the rectangle, and λ_j is the corresponding eigenvalue of $J(\theta^i)$.

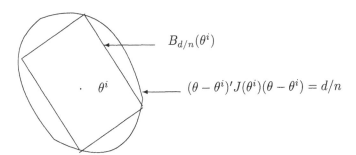

Fig. 6.1. A hyper-rectangle as an equivalence class in the partition.

To show this rotate the coordinates by matrix Q with $Q' = Q^{-1}$ such that in the new coordinate system $QJQ' = \Lambda$, where Λ is a diagonal matrix of the elements given by the eigenvalues of J. Then in the new coordinate system $x'Jx = \sum_j \lambda_j y_j^2$. Let the sides of a rectangle whose corners touch the surface of the hyper ellipsoid, $y'\Lambda y = d/n$, have the lengths $2a_i$. We need to maximize the volume, which can be done by Lagrange's technique:

$$\max_{\{a_i\}} 2^k \prod_i a_i - \lambda \left(\sum_j \lambda_j y_j^2 - d/n\right) .$$

Setting the derivatives to zero gives

$$2^k \prod_{j \neq i} a_j - 2\lambda a_i = 0$$

for all i. Multiply both sides by a_i and sum over all i, which gives the coefficient $\lambda = \frac{kVn}{2d}$. With this then the maximizing sides are given by

$$2a_i = 2\sqrt{d/(kn\lambda_i)}$$

and the volume $2^k \prod_i a_i$, which agrees with (6.3), because the determinants $|J| = |\Lambda|$ are equal.

It is clear that unless the Fisher information matrix is constant we cannot partition the space in terms of the rectangles. The rectangle would have to be curvilinear, as described in [66]. For finite values of n the volume of a curvilinear rectangle no longer is given exactly by (6.3), and we would have to estimate the error. There is another way, which amounts to a piecewise constant approximation of $J(\theta)$. It was used in [3], not for a partition but for a cover of the parameter space in terms of the ellipsoids; see also [46]. This permits an easier way to handle the approximation errors.

Partition first the parameter space into hyper-cubes A_ν of side length r centered at θ^ν. On the boundary the equivalence classes will be portions of the cubes. Each cube is then partitioned into rectangles $B_{d/n}(\theta^i)$, and again on the boundary the equivalence classes are portions of the rectangles. There are $(1 + O(r/\sqrt{d}))r^k/|B_{d/n}(\theta^\nu)|$ rectangles in each cube. When we let n grow and $r = r_n$ shrink at the rate $o(\sqrt{d/n}/r_n)$, the piecewise constant approximation error will go to zero, and for large n we get the desired partition Π_n to a good approximation. We mention that we do not need to construct such a partition. It will be enough to know that one exists.

6.2 Model Complexity

A starting point is the normalized maximum-likelihood (*NML*) model, rewritten here as

$$\hat{f}(x^n; \mathcal{M}_k) = \frac{f(x^n; \hat{\theta}(x^n))}{C_{k,n}} \tag{6.4}$$

$$C_{k,n} = \int_{\hat{\theta}(y^n) \in \Omega^\circ} f(y^n; \hat{\theta}(y^n)) dy^n \tag{6.5}$$

$$= \int_{\hat{\theta} \in \Omega^\circ} g(\hat{\theta}; \hat{\theta}) d\hat{\theta} , \tag{6.6}$$

where Ω° is the interior of a parameter space Ω, and $g(\hat{\theta}; \theta)$ is the density function on the statistic $\hat{\theta}(y^n)$ induced by $f(y^n; \theta)$. The notations dy^n and

$d\hat{\theta}$ refer to differential volumes. The integrals are sums for discrete data, and $g(\hat{\theta}; \theta)$ is a probability mass function.

By the fact that the rectangles partition the parameter space we can write (ignoring the fact that the equivalence classes are portions of the rectangles on the boundary)

$$C_{k,n} = \sum_i Q_{d/n}(i) ,$$

where

$$Q_{d/n}(i) = \int_{\hat{\theta} \in B_{d/n}(i)} g(\hat{\theta}; \theta) d\hat{\theta}; \tag{6.7}$$

we also wrote $B_{d/n}(\theta^i) = B_{d/n}(i)$. We obtain a discrete prior for the rectangles,

$$w_{d/n}(\theta^i) = \frac{Q_{d/n}(i)}{C_{k,n}} . \tag{6.8}$$

Assuming the continuity of $J(\hat{\theta})$ and hence of $g(\hat{\theta}; \theta)$, we have by (6.3)

$$Q_{d/n}(i) \to g(\hat{\theta}; \theta)|B_{d/n}(i)| \to \left(\frac{2d}{k\pi}\right)^{k/2} . \tag{6.9}$$

Since the right-hand side of (6.8) does not depend on θ^i asymptotically, we have an asymptotically uniform prior, and the model $f(y^n; \theta^i)$ can be encoded with the code length

$$L(\theta^i) = \log C_{k,n} - \log Q_{d/n}(i) \tag{6.10}$$

$$\to \log C_{k,n} + \frac{k}{2} \log \frac{k\pi}{2d} . \tag{6.11}$$

It is clear that this code length is asymptotically optimal in the sense of the worst-case model.

To conclude this section we make a connection between the asymptotically uniform discrete prior and Jeffreys' prior $\pi(\hat{\theta})$ (5.24),

$$\frac{w_{d/n}(\theta^i)}{|B_{d/n}(\theta^i)|} \to \pi(\hat{\theta}) .$$

What this means is that neither the canonical prior nor Jeffreys' prior are "noninformative" in the Bayesian sense. However, they are noninformative in the sense that the discrete prior $w_{d/n}(\theta^i)$ is asymptotically uniform and hence absolutely noninformative. We can also write

$$\sum_i f(x^n; \theta^i) w_{d/n}(\theta^i) \to f_\pi(x^n) = \int_\Omega f(x^n; \theta) \pi(\theta) d\theta , \tag{6.12}$$

as $d \to 0$, which again shows in a different way the intimate relationship between the mixture universal model and the *NML* model

$$\hat{f}(x^n; \mathcal{M}_\gamma) = \sum_i w_{d/n}(\theta^i) \hat{f}(x^n|\theta^i) = \frac{1}{C_{\gamma,n}} \sum_i \hat{f}(x^n|\theta^i) . \tag{6.13}$$

6.3 Sets of Typical Sequences

How should we replace the set A, formalizing the properties of the string x^n in the algorithmic theory? The intent there is that all the strings in A are equal, sharing the properties specified by A, and hence being "typical" of this set. This suggests that we should replace set A by a set of "typical" strings of the model $f(y^n; \theta^i)$ – namely, a set of strings for which $\hat{\theta}(y^n) \in B_{d/n}(i)$ for some d. Then the complexity of this set will be the code length $L(\theta^i)$ needed to describe θ^i.

Just as $\log |A|$ is the code length of the worst-case sequence in A, we need the code length of the worst-case sequence y^n such that $\hat{\theta}(y^n) \in B_{d/n}(i)$. It is obtained by Taylor series expansion as follows:

$$- \ln f(y^n; \theta^i) = - \ln f(y^n; \hat{\theta}(y^n)) + \frac{1}{2}d \,, \tag{6.14}$$

where y^n denotes a sequence for which

$$n(\hat{\theta}(y^n) - \theta^i)' \hat{J}(y^n, \tilde{\theta}^i)(\hat{\theta}(y^n) - \theta^i) = d \,.$$

Here $\hat{J}(y^n, \hat{\theta})$ is the empirical Fisher information matrix, (5.27), rewritten here as

$$\hat{J}(y^n, \hat{\theta}) = -n^{-1} \left\{ \frac{\partial^2 \ln f(y^n; \hat{\theta})}{\partial \hat{\theta}_j \partial \hat{\theta}_k} \right\} , \tag{6.15}$$

and $\tilde{\theta}^i$ is a point between θ^i and $\hat{\theta}(y^n)$. We also assume that $\hat{J}(y^n, \hat{\theta}(y^n))$ converges to $\hat{J}(y^n, \hat{\theta})$ as $\hat{\theta}(y^n) \to \hat{\theta}$. We note that for the exponential family $\hat{J}(y^n, \hat{\theta}(y^n)) = J(\hat{\theta}(y^n))$.

Since replacing $- \ln f(y^n; \hat{\theta}(y^n))$ by $- \ln f(x^n; \hat{\theta}(x^n))$ for both $\hat{\theta}(y^n)$ and $\hat{\theta}(x^n)$ in the same rectangle results in an insignificant error, we define the analog of Kolmogorov's structure function $h_{x^n}(\alpha)$ as follows:

$$h_{x^n}(\alpha) = \min_d \left\{ - \ln f(x^n; \hat{\theta}(x^n)) + \frac{1}{2}d : L(\theta^i) \le \alpha \right\} . \tag{6.16}$$

For the minimizing d the inequality will have to be satisfied with equality,

$$\alpha = \frac{k}{2} \ln \frac{\pi k}{2d} + \ln C_n \,, \tag{6.17}$$

and with the asymptotic approximations we get

$$d_\alpha = \frac{\pi k}{2} C_n^{2/k} e^{-2\alpha/k} \,. \tag{6.18}$$

From (6.14),

$$h_{x^n}(\alpha) = - \ln f(x^n; \hat{\theta}(x^n)) + d_\alpha/2 \,. \tag{6.19}$$

With this we get further

$$h_{x^n}(\alpha) + \alpha = -\ln \hat{f}(x^n) + d_\alpha/2 - \frac{k}{2}\ln d_\alpha + \frac{k}{2}\ln \frac{\pi k}{2} . \qquad (6.20)$$

Recall Equation (4.6), rewritten here as

$$\min\{\alpha : h_{x^n}(\alpha) + \alpha \doteq K(x^n)\} , \qquad (6.21)$$

which gives the crucial and optimal way to separate the complexity of the data into the complexity $\bar{\alpha}$ of the model and the complexity $h_{x^n}(\bar{\alpha})$ of the "noise", as it were. However, we do not want to ignore constants, which means that instead of the approximate equality $\doteq$, we want to write an equality. This poses a bit of a problem, because for no value of α could we reach the stochastic complexity $-\log \hat{f}(x^n)$. We resolve this problem by asking for the smallest value of α for which the two-part complexity on the left-hand side reaches its minimum value, which amounts to the *MDL* principle. In our case then we ask for the smallest, and in fact the only, value for α that minimizes the two-part code length (6.20),

$$\min_\alpha\{h_{x^n}(\alpha) + \alpha\} = -\ln f(x^n; \hat{\theta}(x^n)) + \frac{k}{2} + \frac{k}{2}\ln \frac{\pi}{2} + \ln C_{k,n} . \qquad (6.22)$$

The minimum is reached for $d_\alpha = k$ at the point

$$\bar{\alpha} = \frac{k}{2}\ln \frac{\pi}{2} + \ln C_{k,n} . \qquad (6.23)$$

These are illustrated by the following graph (Figure 6.2).

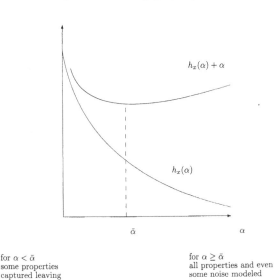

$h_x(\alpha) + \alpha$

$h_x(\alpha)$

$\bar{\alpha}$ α

for $\alpha < \bar{\alpha}$
some properties
captured leaving
a lot as noise

for $\alpha \geq \bar{\alpha}$
all properties and even
some noise modeled

Fig. 6.2. A schematic representation of the structure function and the optimal model.

We consider (6.22) as a universal sufficient statistics decomposition, in which

- the logarithm of the number of optimal models, (6.23), represents the amount of *learnable information* in the data that can be extracted with the model class considered, while the rest,
- the optimized structure function consisting of the first two terms in (6.22), represents the amount of "noise" that cannot be explained with the optimal model.

For large n the logarithm of the number of optimal models is not much larger than the number obtained by Balasubramanian, namely, $\ln C_{k,n}$, which corresponds to the value $\bar{d} = (k\pi/2)$, (6.11). The corresponding *universal sufficient statistics* decomposition is then

$$- \ln f(x^n; \hat{\theta}(x^n)) + \frac{k\pi}{4} + \ln C_{k,n} .$$

Next, consider the line, analogous to the *sufficiency line* in the algorithmic theory [79],

$$L(\alpha) = - \ln \hat{f}(x^n) + \frac{k}{2} \ln \frac{\pi e}{2} - \alpha . \tag{6.24}$$

By (6.20) and (6.22), the curve $h_{x^n}(\alpha)$ lies above the line $L(\alpha)$ for $0 \le \alpha \le \alpha_{max}$, where

$$\alpha_{max} = - \ln \hat{f}(y^n) + \frac{k}{2} \ln \frac{\pi e}{2} ,$$

except for the point $\bar{\alpha}$, where the line is its tangent. Indeed,

$$\frac{dh_{x^n}(\alpha)}{d\alpha} = - \frac{d\alpha}{k} .$$

The universal sufficient statistics decomposition immediately resolves the puzzling anomaly in the traditional statistics that the maximum-likelihood estimate $\hat{\theta}(x^n)$ of the values of the parameters is acceptable, but the same estimate of their number $\hat{k}(x^n)$ is not. Yet both are just parameters, and one and the same principle of estimation should be applicable to both. The explanation is that the maximum-likelihood principle should be rejected in both cases, because we are not then distinguishing between noise and the learnable part, i.e., the information. In case of $\hat{\theta}(x^n)$ the damage is minor for large amounts of data, because of the convergence properties of these estimates. However, in case of $\hat{k}(x^n)$ the damage is devastating. In light of the decomposition above we now see that in every case we should separate the noise part from the information and fit parameters only such that we capture the information – in other words, to the precision given by the quantization defined by the cells $B_{\hat{d}/n}(i)$.

Example 1. We derive the universal sufficient statistics decomposition for the Bernoulli class $\mathcal{B}$, with $P(x = 0) = \theta$ as the parameter. The *NML* distribution is given by

$$\hat{P}(x^n) = \frac{P(x^n; \hat{\theta}(x^n))}{\sum_m \binom{n}{m}(\frac{m}{n})^m (\frac{n-m}{n})^{n-m}} , \tag{6.25}$$

where $\hat{\theta} = \hat{\theta}(x^n) = n_0(x^n)/n$, $n_0(x^n)$, denoting the number of 0's in x^n, and $P(x^n; \hat{\theta}(x^n)) = \hat{\theta}^{n_0(x^n)}(1 - \hat{\theta})^{n-n_0(x^n)}$. Write this in the form

$$\hat{P}(x^n) = \frac{1}{\binom{n}{n_0(x^n)}} \times \pi_n(\hat{\theta}(x^n)) , \tag{6.26}$$

where

$$\pi_n(\hat{\theta}(x^n)) = \frac{\binom{n}{n_0(x^n)} P(x^n; \hat{\theta}(x^n))}{\sum_m \binom{n}{m}(\frac{m}{n})^m (\frac{n-m}{n})^{n-m}} . \tag{6.27}$$

In order to evaluate the resulting ideal code length $-\ln \hat{P}(x^n)$, we use the important and ubiquitous Stirling's approximation formula in the form refined by Robbins:

$$\ln n! = (n + 1/2) \ln n - n + \ln \sqrt{2\pi} + R(n) , \tag{6.28}$$

where

$$\frac{1}{12(n + 1)} \le R(n) \le \frac{1}{12n} .$$

This permits us to evaluate the terms in the sum in Equation 6.27 to a sufficient accuracy for us as –

$$\ln \binom{n}{m} \cong nh(m/n) - \frac{1}{2} \ln[(m/n)(n - m)/n] - \frac{1}{2} \ln n - \ln \sqrt{2\pi} ,$$

where $h(p)$ is the binary entropy at p. This gives

$$\binom{n}{m}\left(\frac{m}{n}\right)^m \left(\frac{n-m}{n}\right)^{n-m} \cong \frac{1}{\sqrt{2\pi n}}[(m/n)(n - m)/n]^{-1/2} .$$

Recognizing the sum in the denominator of Equation (6.27) (with step length $1/n$ rather than $1/\sqrt{n}$) as an approximation of a Riemann integral, we get it approximately as

$$\sqrt{n/(2\pi)} \int_0^1 \frac{1}{\sqrt{p(1 - p)}} dp = \sqrt{n\pi/2} ,$$

where the integral of the square root of the Fisher information $J(p) = 1/(p(1 - p))$ is a Dirichlet integral with the value π. Finally,

$$-\log \hat{P}(x^n) = nh(n_0/n) + \frac{1}{2} \log \frac{n\pi}{2} + o(1) , \tag{6.29}$$

where the normalizing coefficient is given by

$$\log C_{1,n} = \frac{1}{2} \log \frac{n\pi}{2} + o(1) . \tag{6.30}$$

The added term $o(1)$, which goes to zero as $n \to \infty$, takes care of the errors made by the application of Stirling's formula. More accurate approximations for the normalizing coefficient exist; see, for instance, [72].

The width of the equivalence class for the parameter d is by (6.3)

$$|B_{d/n}(i)| = \left(\frac{4d}{n}\right)^{1/2} ((i/n)(1 - i/n))^{1/2} , \tag{6.31}$$

from which the optimal interval length is obtained for $d = 1$. By the approximations made the index i of the center should not be too close to zero nor unity.

The probability distribution for the centers of the optimal equivalence classes is from (6.8) and (6.11)

$$w_{1/n}(i) = \frac{\sqrt{2/\pi}}{C_{1,n}} .$$

The universal sufficient statistic decomposition (6.22) is then given by

$$nh(n_1/n) + \frac{1}{2} + \frac{1}{2} \log n + \log \frac{\pi}{2} , \tag{6.32}$$

where n_1 denotes the number of 1's in the string x^n and $h(n_1/n)$ is the binary entropy function evaluated at the point n_1/n.

Historical Notes

As stated before, this development of the structure function was inspired by Kolmogorov's work, which I learned from [79]. Although I have had the main concepts I needed for a fairly long time, some of the intricate details have been proved only recently in conjunction with the topic in the next chapter [63].

7

Optimally Distinguishable Models

In [3] and [4] Balasubramanian showed the most interesting result that $C_{k,n}$, also known as the Riemann volume, for a fixed γ with k parameters gives the maximal number of *distinguishable* models that can be obtained from data of size n. His idea of distinguishability is based on differential geometry and is somewhat intricate, mainly because it is defined in terms of covers of the parameter space rather than partitions. The fact is that any two models f_i and f_j, no matter how far apart, will have intersecting supports, and hence the sense in which two models can be distinguished does not have a direct interpretation. We wish to refine Balasubramanian's idea by obtaining a measure of the separation between the models, which is both intuitively appealing and also can be easily interpreted. Moreover, importantly, we show that there is a value for the parameter d and the number of the equivalence classes $B_{d,n}(\theta^i)$ for which this measure is optimized.

Considering the partition Π_n of the rectangles in the previous section, we construct from $f(y^n; \hat{\theta}(y^n))$ a special model, defined by the center θ^i of the rectangle $B_{d/n}(i) = B_{d/n}(\theta^i)$ and having the same rectangle as the support

$$\hat{f}(y^n|\theta^i) = \begin{cases} f(y^n; \hat{\theta}(y^n))/Q_{d/n}(i) & \text{if } \hat{\theta}(y^n) \in B_{d/n}(i), \\ 0 & \text{otherwise} \end{cases} \qquad (7.1)$$

where $Q_{d/n}(i)$ was given in (6.9).

To see that this is a well-defined model, note that

$$f(y^n; \theta) = f(y^n, \hat{\theta}(y^n); \theta) = f(y^n|\hat{\theta}(y^n); \theta)g(\hat{\theta}(y^n); \theta) .$$

By integrating first over y^n such that $\hat{\theta}(y^n) = \hat{\theta}$ and then over $\hat{\theta} \in B_{d/n}(i)$, we get

$$\int_{\hat{\theta}(y^n) \in B_{d/n}(i)} f(y^n; \hat{\theta}(y^n)) dy^n = \int_{\hat{\theta} \in B_{d/n}(i)} g(\hat{\theta}; \hat{\theta}) d\hat{\theta}.$$

While these models are hardly useful for statistical applications, they have the distinct property of being perfectly distinguishable, because their supports are disjoint. Indeed, these could not be applied to prediction, because we could not be certain that $\hat{\theta}(y^{n+1})$ falls in the support $B_{d/n}(i)$. Our main objective is to study the models $\hat{f}(y^n; \theta^i)$, the support of each being the set of all sequences y^n. These undoubtedly are the natural models defined by the parameters, which can be used to model future data y_t, $t > n$ by $f(y_t|x^n; \theta^i(x^n))$. Consider the KL distance $D(\hat{f}(Y^n|\theta^i)\|f(Y^n; \theta^i))$. The smaller this distance is the closer the real model $\hat{f}(y^n; \theta^i)$ is to the perfectly distinguishable model $\hat{f}(y^n|\theta^i)$, and if we minimize the distance we get a measure of how well adjacent models can be distinguished:

$$\min_d D(\hat{f}(Y^n|\theta^i)\|f(Y^n; \theta^i)) . \tag{7.2}$$

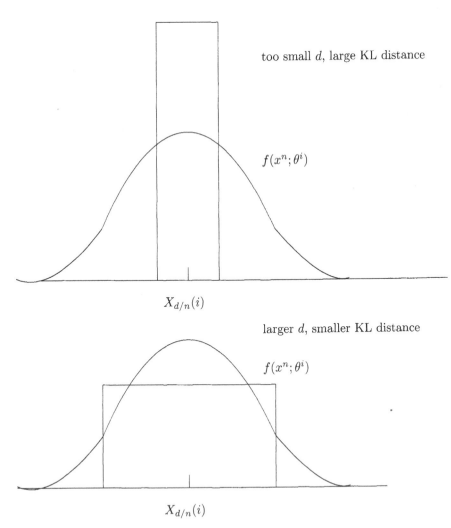

Fig. 7.1. Illustration of the dependence of KL distance on d.

We show next that there is a meaningful minimum with the assumption we have made, namely, that

$$g(\hat{\theta};\hat{\theta})n^{-k/2} \rightarrow \frac{|J(\hat{\theta})|^{1/2}|}{(2\pi)^{k/2}} \tag{7.3}$$

together with the simplifying assumptions made about the empirical Fisher information matrix. A detailed derivation without (7.3) is made in [63].

We have again by Taylor's expansion

$$D(\hat{f}(Y^n|\theta^i)\|f(Y^n;\theta^i)) = \frac{1}{Q_{d/n}(i)} \int_{y^n:\hat{\theta}(y^n)\in B_{d/n}(i)} f(y^n;\hat{\theta}(y^n))[\ln 1/Q_{d/n}(i)$$
$$+ \frac{n}{2}\xi(y^n)'\hat{J}(y^n,\tilde{\theta}(y^n))\xi(y^n)]dy^n, \tag{7.4}$$

where $\xi(y^n) = \hat{\theta}(y^n) - \theta^i$ and $\tilde{\theta}(y^n)$ between θ^i and $\hat{\theta}(y^n)$. Integration over the first term gives just $\ln 1/Q_{d/n}(i)$. With the simplifying assumptions made we can integrate first over y^n such that $\hat{\theta}(y^n) = \hat{\theta}$ and then over $\hat{\theta}$. The integral of the second term becomes with some wrestling, which we omit (see [63]),

$$\int_{\hat{\theta}\in B_{d/n}(i)} \frac{g(\hat{\theta};\hat{\theta})}{Q_{d/n}(i)}(\hat{\theta}-\theta^i)'J(\theta^i)(\hat{\theta}-\theta^i)d\hat{\theta}$$
$$\cong \frac{1}{|B_{d/n}(i)|} \int_{\hat{\theta}\in B_{d/n}(i)} (\hat{\theta}-\theta^i)'J(\theta^i)(\hat{\theta}-\theta^i)d\hat{\theta} . \tag{7.5}$$

Next we evaluate the right-hand side integral. By rotation of the coordinates

$$\xi'J(\theta^i)\xi = \sum_{j=1}^{k}\lambda_j\xi_j^2 ,$$

where $\{\lambda_j\}$ is the set of the eigenvalues of $J(\theta^i)$, and

$$\int_{\theta\in B_{d/n}(i)} (\theta-\theta^i)'J(\theta^i)(\theta-\theta^i)d\theta = 2^k \int_0^{\mu_1}\cdots\int_0^{\mu_k}\prod_{j=1}^{k}d\xi_j\sum_{j=1}^{k}\lambda_j\xi_j^2$$
$$= \frac{2^k}{3}\sum_j\mu_j^3\prod_{s\neq j}\mu_s = \frac{d}{3n}|B_{d/n}(i)| ,$$

where $2\mu_j = 2\sqrt{d}/\sqrt{k\lambda_j}$ is the j'-th side length of the rectangle.

Then with (6.9)

$$D(\hat{f}(Y^n|\theta^i)\|f(Y^n;\theta^i)) \rightarrow \frac{k}{2}\ln\frac{k\pi}{2d} + d/6 . \tag{7.6}$$

The right-hand side is minimized at $\hat{d} = 3k$, and the minimum value of the KL distance is

$$\min_{d} D(\hat{f}(Y^n|\theta^i)\|f(Y^n;\theta^i)) \rightarrow \frac{k}{2}\ln\frac{\pi e}{6} \ .$$

To conclude this section, we establish a link with the structure function and get the second interpretation of optimal distinguishability. Instead of calculating the structure function in (6.16) with the worst case code length in the set $B_{d/n}(i)$, given the model $f(y^n;\theta^i)$, as

$$L(x^n|\theta^i) = \max_{y^n}\log\frac{f(y^n;\hat{\theta}(y^n))}{f(y^n;\theta^i)} = d/2 \ ,$$

take the average

$$\frac{1}{|B_{d/n}(i)|}\int_{\hat{\theta}\in B_{d/n}(i)}\log\frac{f(y^n;\hat{\theta}(y^n))}{f(y^n;\theta^i)}dy^n \rightarrow d/6 \ ,$$

which gives the structure function

$$h_{x^n}(\alpha) = \min_{\alpha}\{-\ln f(x^n;\hat{\theta}(x^n))$$

$$+ \frac{1}{|B_{d/n}(i)|}\int_{\hat{\theta}\in B_{d/n}(i)}\log\frac{f(y^n;\hat{\theta}(y^n))}{f(y^n;\theta^i)}dy^n : L(\theta^i) \le \alpha\} \ . \ (7.7)$$

With (6.11) we get the minimizing value for $d = d_\alpha$ as $\hat{d}$, which satisfies $\hat{d} \rightarrow 3k$ in agreement with the value minimizing (7.6). We then have the second (asymptotic) interpretation of optimal distinguishability: *It represents an optimal balance between the uncertainties due to sampling, as measured by the average of the variance $(\theta-\theta^i)'J(\theta^i)(\theta-\theta^i)$, and the uncertainty in finding the optimally quantized model $f(y^n;\theta^i)$, as represented by the code length for θ^i*. This is particularly relevant in hypothesis testing, which we discuss below in Chapter 9: Applications. It is clear that the optimal hyper-rectangles give in a natural way what traditionally are meant by "confidence intervals".

7.1 Special Cases

In the two special cases, the Bernoulli distributions and the normal ones, the central limit theorem is not needed, and we can calculate the KL distance and its minimum exactly.

7.1.1 Bernoulli class

Let $P(X = 0) = p$ and $\hat{p} = m_y/n$, where m_y is the number of 0's in y^n. We have

$$g(\hat{p}(y^n);\hat{p}(y^n)) = \binom{n}{m_y}(m_y/n)^{m_y}(1 - m_y/n)^{n-m_y} \ .$$

Consider the interval centered at i/n, whose length is

$$|B_{d/n}(i/n)| = \left(\frac{4d}{n}\right)^{1/2} ((i/n)(1-i/n))^{1/2} , \tag{7.8}$$

and for m an integer compute the sum

$$Q_{d/n}(i/n) = \sum_{m/n \in B_{d/n}(i/n)} \binom{n}{m} (m/n)^m (1-m/n)^{n-m} .$$

Then find the value of d that minimizes the KL distance

$$\min_d \left\{ \log 1/Q_{d/n}(i) + \frac{1}{Q_{d/n}(i/n)} \right.$$

$$\left. \times \sum_{m/n \in B_{d/n}(i/n)} g\left(\frac{m}{n}; \frac{m}{n}\right) \log \frac{(m/n)^m (1-m/n)^{n-m}}{(i/n)^m (1-i/n)^{n-m}} \right\} .$$

Clearly this can be done easily unless n is too large. Asymptotically we have $\hat{d} = 3$.

7.1.2 Normal distributions

For the class of normal distributions with two parameters, the mean μ and variance τ, the asymptotic value for the parameter giving the optimal partition is $\hat{d} = 6$. Since the central limit theorem is not needed in this case, we could calculate the KL distance exactly and perform the minimization in (7.2). However, we get a quite accurate evaluation of $\hat{d}$ more easily by making simple approximations to the integrals involved.

We begin by calculating the density function $g(\hat{\theta}; \theta)$, evaluated at the ML estimates $\theta = \hat{\theta} = \hat{\mu}, \hat{\tau}$,

$$\hat{\mu} = \frac{1}{n} \sum_{t=1}^{n} y_t$$

$$\hat{\tau} = \frac{1}{n} \sum_{t=1}^{n} (y_t - \hat{\mu})^2 .$$

The estimate $\hat{\mu}$ has the normal distribution with mean μ and variance τ/n, while $n\hat{\tau}/\tau$ has the chi-square distribution of $n-1$ degrees of freedom; see also (9.22) below for $k = 1$. By applying Stirling's approximation to the gamma function in the chi-square distribution and ignoring the slight error term, which does not change the final outcome, we get

$$g(\hat{\theta}; \hat{\theta}) = \frac{\sqrt{n}}{\sqrt{2\pi\hat{\tau}}} n^{\frac{n-1}{2}-1} (2e)^{-n/2} \sqrt{2}\Gamma^{-1}\left(\frac{n-1}{2}\right) \tag{7.9}$$

$$= A_n n \hat{\tau}^{-3/2} , \tag{7.10}$$

where

$$A_n = \frac{1}{2\pi} \left(\frac{2}{e}\right)^{3/2} \left(\frac{n}{n-3}\right)^{(n-4)/2} .$$

We calculate next the Fisher information matrix. Differentiating

$$-\ln f(y^n; \mu, \tau) = \frac{n}{2} \ln(2\pi\tau) + \frac{1}{2\tau} \sum_t (y_t - \mu)^2$$

twice with respect to μ and τ and taking the mean, we get

$$J(\theta) = \begin{bmatrix} 1/\tau & 0 \\ 0 & 1/\tau^2 \end{bmatrix} ,$$

and the Fisher information itself is given by $|J(\theta)| = \tau^{-3}$.

Consider the two-dimensional rectangle $B_{d/n}(i)$ centered at the point $\theta^i = (\mu, \tau)$ of the side lengths $\tau_2 - \tau_1 = 2\sqrt{2d/n}\tau$ and $\mu_2 - \mu_1 = 2\sqrt{\tau d/n}$. Next, calculate the integral $Q_{d/n}(i)$ in (6.9):

$$Q_{d/n}(i) = 2A_n n \sqrt{\frac{\tau d}{n}} \int_{\tau_1}^{\tau_2} \hat{\tau}^{-3/2} d\hat{\tau}$$

$$= 4A_n \sqrt{nd} \frac{\sqrt{1 + \sqrt{2d/n}} - \sqrt{1 - \sqrt{2d/n}}}{\sqrt{1 - 2d/n}} \tag{7.11}$$

$$= 4\sqrt{2} A_n d[1 - d/n + O((d/n)^2)] . \tag{7.12}$$

The main task is to calculate the Kullback–Leibler distance in (7.4):

$$D(\hat{f}(Y^n|\theta^i) \| f(Y^n; \theta^i)) = \frac{1}{Q_{d/n}(i)} \int_{y^n : \hat{\theta}(y^n) \in B_{d/n}(i)} f(y^n; \hat{\theta}(y^n))$$

$$\times \ln \frac{f(y^n; \hat{\theta}(y^n))}{Q_{d/n}(i) f(y^n; \theta^i))} dy^n \tag{7.13}$$

$$= -\ln Q_{d/n}(i) + \frac{A_n n^2}{2Q_{d/n}(i)}$$

$$\int_{\hat{\theta} \in B_{d/n}(i)} \hat{\tau}^{-3/2} \left[\frac{\hat{\tau} - \tau}{\tau} - \ln \frac{\hat{\tau}}{\tau} + \frac{1}{\tau}(\hat{\mu} - \mu)^2 \right] d\hat{\tau} d\hat{\mu} ,$$

$$\tag{7.14}$$

where we used the identity

$$\sum_t (y_t - \mu)^2 = \sum_t [(y_t - \hat{\mu}) + (\hat{\mu} - \mu)]^2 = n[\hat{\tau} + (\hat{\mu} - \mu)^2] ,$$

and integrated first over y^n such that $\hat{\theta}(y^n) = \hat{\theta}$ and then over $\hat{\theta}$. It is possible to calculate the four integrals needed exactly, but we get more information from an easy approximation. Writing

$$\frac{\hat{\tau}}{\tau} = \frac{\hat{\tau} - \tau}{\tau} + 1$$

and applying the expansion $\ln(1 + \epsilon) = \epsilon + O(\epsilon^2)$, we get the expression in brackets as follows:

$$\frac{1}{\tau}(\hat{\mu} - \mu)^2 + O(2d/n).$$

The integral of the square with respect to $\hat{\mu}$ is then given by

$$I_1 = \int_{\mu_1}^{\mu_2} (\hat{\mu} - \mu)^2 d\hat{\mu} = \frac{2}{3}\left(\frac{\tau d}{n}\right)^{3/2}.$$

With the integral

$$I_2 = \int_{\tau_1}^{\tau_2} \hat{\tau}^{-3/2} d\hat{\tau} = 2\sqrt{\frac{2d}{n\tau}}, \tag{7.15}$$

where we used the substitutions

$$\tau_1 = \tau(1 - \sqrt{2d/n})$$
$$\tau_2 = \tau(1 + \sqrt{2d/n}),$$

we then get from (7.14) and (7.12)

$$D(\hat{f}(Y^n|\theta^i)\|f(Y^n;\theta^i)) = d/6 - \ln d + d/n + O(1/n^2). \tag{7.16}$$

The minimizing value $\hat{d} = 6 - 36/n + O(1/n^2)$ modifies slightly the earlier general asymptotic value $\hat{d} = 6$.

Historical notes

Although it has been clear from the very beginning of the MDL principle that the optimal model calls for quantization of the parameter values to the precision $O(1/\sqrt{n})$ per parameter, which also limits their precision by the Cramer–Rao inequality such that one cannot estimate them more accurately. However, it was only through Balasubramanian's work [3] that I was able to sharpen the arguments to the current, in my mind fundamental, notion of *optimal distinguishability* with its far-reaching implications.

8

The *MDL* Principle

We have mentioned the *MDL* principle on several occasions somewhat loosely as the principle that calls for finding the model and model class with which the data together with the model and model class, respectively, can be encoded with the shortest code length. Actually to apply the principle we must distinguish between two types of models – those for data compression and others for general statistical purposes such as prediction. In data compression, we apply the models to the same data from which the models are determined. Hence these models need not have any predictive power; and, in fact, to get the shortest code length we do not even need to fit models in the class considered, say, $\mathcal{M}_\gamma$. This is because the universal *NML* model gives a code length, which we called the stochastic complexity and which we consider to be the shortest for all intents and purposes.

In the early days of control theory the naive thinking was that most processes to be controlled are linear, and the task at hand is to "identify" a linear system from its impulse response, with perhaps some gaussian noise added. Similarly, in statistics the thinking has been advanced that the main problem of statistics is to "identify" a distribution from the data it creates by sampling as if the world were made of random variables. In reality, the problems of statistics are much more severe. The main task is to find constraints that restrict the observed data, so that thus amount to a model of the data. All restrictions in the data affect the code length with which we can encode the data: the less space they allow for the data, the shorter the coding that can be achieved, which gives us the means both to measure the strength of constraints and to find them, which provides the foundation for the *MDL* principle in modeling.

By the *MDL* principle we seek to minimize a two-part code length: the first part giving the code length for the model and the second the code length for the data. Since we are dealing with probability models we, equivalently, need to maximize the joint probability for the data and the model. This makes sense, for we need to calculate the probability of the coincidence of the fixed data and the different models – i.e., the joint probability – and pick the maximizing model. Whatever way we use to encode the models, coding of the data, given a model, must be as efficient as we can make it.

If we wish to learn the optimal model class, say, $\mathcal{M}_\gamma$, where $\gamma \in \Gamma$, by the *MDL* principle, we need to minimize the two-part code length

$$\min_{\gamma, L \in \mathcal{L}} \{\log 1/\hat{f}(x^n; \mathcal{M}_\gamma) + L(\gamma)\} , \qquad (8.1)$$

where $\mathcal{L}$ denotes a family of code lengths of decodable codes for γ. If the set Γ of the structures is small, we may use the uniform prior, which gives the code length $L(\gamma) = \log |\Gamma|$ for all $\gamma \in \Gamma$, and we may ignore this as well as the class $\mathcal{L}$, because there is no point in trying to find a more efficient coding of γ. In such a case (8.1) is equivalent with the posterior maximization, although the posterior is not the same as in the Bayesian procedure. In general, however, the class $\mathcal{L}$ is needed and a shorter code length $L(\hat{\gamma})$ for the optimizing structure than the uniform can be found, which makes the criterion (8.1) different from the posterior maximization one.

A frequently occurring special case where this happens is when the structure amounts to the number of parameters, and since we do not wish to fit more parameters than the number of data points, n, we have an upper bound. This may be given and fixed or not. If it is given we may take the prior for k either $1/n$ with the code length $\log n$, which can be ignored, or the conditional $w(k|n)$ in (2.9) with the code length

$$L(k|n) = \log k + \log \ln(en) ,$$

or we may pick in the criterion (8.1) the shorter of the two. It is clear that we could construct several other code lengths similar to $L(k|n)$, but they are unlikely to make a noticeable improvement. If n is not given, we take $L(k)$ as the unconditional code length for the integers,

$$\log 1/w^*(k) = \log k + \log \log k + \ldots + \log c ,$$

as discussed in the section for coding of the integers, where the constant c is about 2.865, and where only positive terms in the iterated logarithms are included.

Another important model class where $L(\gamma)$ cannot be ignored arises in regression problems, where the structures consist of subsets of $\{1, \ldots, n\}$. Since there are 2^n subsets, one candidate prior is $w(\gamma) = 1/2^n$ with the code length n, which is likely to be a large number. Another, suggested by T. Roos [67] is $w(\gamma|k) = \binom{n}{k}^{-1}$, which requires additional encoding of k with the code length $L(k) = \min\{\log n, \log k + \log \ln(en)\}$.

Returning to the problem (8.1) we conclude that the minimizing $\hat{L}(\hat{\gamma})$ gives the amount of *structure information* that can be learned from Γ. All the strings y^n such that $\hat{\gamma}(y^n) = \hat{\gamma}(x^n)$ are equal, giving us the same amount of structure information.

If we want to find the optimal model in the class $\mathcal{M}_{\hat{\gamma}}$, a problem arises, because the set of models has the cardinality of continuum. This means that we must quantize the set of parameters Ω. As explained in Chapter 7, the partition of the parameter space depends on a parameter d, which gives us the code length $L_d(\theta^i)$, (6.11), for the quantized parameters and the corresponding models $f(x^n; \theta^i)$. We regard the strings $\{y^n : \hat{\theta}(y^n) \in B_{d/n}(\theta^i)\}$ as equivalent since all of them specify the same quantized parameter value. Depending on how we calculate the code length of these equivalent strings we get slightly different structure functions and two-part code lengths to minimize

$$\min_d \{\log 1/f(x^n; \hat{\theta}(x^n)) + \rho(d) + L_d(\theta^i)\} , \qquad (8.2)$$

where $\rho(d) = d/2$ if we take the worst case code length in the $B_{d/n}(\theta^i)$ (6.16) and $\rho(d) \cong d/6$ if we take the average. With the minimizing value $\hat{d}$ we can state again that $L_{\hat{d}}(\theta^i)$ (6.11) represents the amount of *information* in the data that can be learned with the optimal model and the model class, and the two first terms represent the amount of *noise* that cannot be explained – i.e., that looks random in light of the model class considered.

The quantization of the real-valued parameters was made out of necessity, although it is also required by optimality. In modeling problems for denoising, the set of structures Γ can be very large indeed (see Section Linear Regression below), and it may happen that many of the structures near the optimum are almost as good as the optimum. But then the minimization in (8.1) may not produce the *MDL* solution after all, for we may get a shorter code length if we partition the set Γ optimally into equal-size subsets. Then the code length $\hat{L}(\hat{\gamma})$ can be shortened without a corresponding increase in the first term. Consider the simple case where each structure is specified by the number of parameters k. Partition the set $\Gamma = \{1, 2, \ldots, n\}$ into discrete intervals $\{I_j : j = 1, 2, \ldots, \lceil n/r \rceil\}$ of length r. Consider the minimization problem

$$\min_r \{\log 1/\hat{f}(x^n; \hat{k}) + \frac{1}{r} \sum_{k \in I_{j(k)}} \log \frac{\hat{f}(x^n; \hat{k})}{\hat{f}(x^n; k)} + \log \lceil n/r \rceil\} , \qquad (8.3)$$

where we wrote $\hat{f}(x^n; \mathcal{M}_k) = \hat{f}(x^n; k)$ and $j(k) = \lceil k/r \rceil$. It's clear that this two-part code length is shorter than or equal to that in (8.1).

The *MDL* principle represents a drastically different foundation for model selection and, in fact, statistical inference in general. It has a number of distinctive features, such as –

- There is no need to assume anything about how the existing data were generated. In particular, unlike in traditional statistics, the data need not be assumed to form a sample from a population with some probability law.
- In this view, the objective in modeling is not to estimate an assumed but "unknown" distribution, be it inside or outside the proposed class of models, but to find good models for the data.

- Most importantly, the principle permits comparison of any two models/model classes, regardless of their type. Hence, it provides a vastly more general criterion than *AIC*, *BIC*, and others that depend only on the number of parameters even for the very special models for which these criteria are applicable.

- Finally, one cannot compress data without taking advantage of the data-restricting features. This means that small random changes in data cannot alter the best-fitting models too much, and the *MDL* models are naturally robust. This is quite evident in the very way the optimally distinguishable models were constructed.

The application of the principle requires, of course, the calculation of the stochastic complexity, which for complex model classes can be a difficult task; and in practice we may have to settle for an upper-bound approximation. The situation, however, is not as bad as in approximating the Kolmogorov complexity, where the upper bounds are noncomputable; i.e., we never know how far we are from the complexity. Here we do have means to assess the amount of the excess; for a good approximation we refer to [46]. A reasonable way to proceed in very complex modeling situations is to decompose the class into smaller ones for which the formulas derived above are valid. The code length required for the links needed to put these together can be estimated, usually by visualizing a concrete coding scheme. The requirement for a prefix code length can be much easier to satisfy than by constructing a prior. An example, given in the introduction, is in the problem of dissecting an image into "scenes" of roughly equal gray level defined by closed curves like loops. There is a chain coding method to describe such loops. It would be a formidable task to construct a prior probability function for the set of all loops on an image.

We would like to mention another important issue related to the restriction of the use of models, for instance, for prediction. The issue has apparently been discussed for the first time in [22]. Since in the *MDL* view no model in general captures all the relevant regular features in the data, we cannot expect to be able to predict reliably just any properties of the data we wish. Rather, we should predict only those properties we have captured by our model. Notice that if the model is regarded as an approximation of an imagined "true" underlying distribution, as is the case in traditional statistics, there are no restrictions on the properties one should predict. To be concrete, suppose we find that the *MDL* model in a class of all Markov models turns out to be a Bernoulli-model, even though the data are generated by a first-order Markov process. However, the prediction $\hat{p}(x^n) = m/n$ of the probability that the symbol is 0, where m denotes the number of zeros in the data string x^n, is still reliable. In fact, even if you had fitted a Markov model of the first order, you would get the same estimate of the probability that the state equals 0. By contrast, the prediction of the probability $P(00) = \hat{p}^2(x^n)$ with the optimal Bernoulli model would be completely unreliable and would differ from the estimate of the relative number of occurrences of two consecutive zeros in the future data that are typical for the correct first order Markov process. In conclusion one can say that the *MDL* models can be used to provide reliable estimates of all the properties captured by the model – even though the models are imperfect.

The three formulas for the shortest code length above work well if the data have properties similar to the requirement that the central limit theorem holds. This is relevant only for large data sets. There are important cases where there are large amounts of data which do not satisfy this requirement, and the formulas do not give the shortest code length as required by the *MDL* principle. How do we know then when this is the case? A clear indication of trouble is if $\hat{k}(x^{n/2}) << \hat{k}(x^n)$, and if the shortest per symbol code length obtained from a large amount of data, namely, from half of them, increases with the second half. In other words, if we have been unable to capture the relevant restrictions by fitting a parametric model as stated. It is for such data that the predictive universal model may be useful.

An example is given in [64], where the number of bins $\hat{m}(x^n)$ in an equal-width histogram was calculated by minimization of the stochastic complexity for data generated by a density function on the unit interval with bounded first derivative. The so-determined number of bins can be shown to behave like $O((n/\log n)^{1/3})$. The stochastic complexity itself is then given by

$$-\frac{1}{n}\log f(x^n) + K\left(\frac{\log n}{n}\right)^{2/3}.$$

When we calculate the code length with the universal predictive manner by taking $\bar{m}(x^t) = \lceil (n^{1/3}/\log n)$, we get the code length as

$$-\frac{1}{n}\log f(x^n) + Kn^{-2/3},$$

which is shorter than the stochastic complexity. Hence, reducing the model cost improves the result. Perhaps this is why Akaike's criterion, which penalizes the model complexity simply by the number of parameters, can take advantage of restrictions in the data which go beyond the capabilities of the individual models in the class $\mathcal{M}$. However, since the criterion has no data-dependent meaning, almost nothing can be explained by Akaike's criterion. We have no idea when it works and why it works when it works. One gets a fully comparable criterion by adding a different multiple of the number of parameters as the penalty term to the negative logarithm of the likelihood function.

In conclusion, it is important to realize that the *MDL* principle has nothing to say about how to select the suggested family of model classes. In fact, this is a problem that cannot be adequately formalized. In practice their selection is based on human judgment and prior knowledge of the kinds of models that have been used in the past, perhaps by other researchers.

Historical notes

Having been aware of the paper by Wallace and Boulton [81] as well as Akaike's work [1], and the algorithmic theory of information, and having done arithmetic coding, I came to realize that the fundamental problem both in data compression and statistics is the modeling problem, and that the code length minimization is the key to both. As a result I published the first paper on the *MDL* principle [50] with a slightly different name. I started developing *MDL* theory further, first alone [51–55, 60], but gradually with others joining in [4, 5, 7, 21–23, 26], to mention a few; and work is still continuing. After all, modeling is about the most fundamental scientific activity there is and is not likely to be finished any time soon.

9

Applications

9.1 Hypothesis Testing

The generally used Neyman–Pearson hypothesis testing is based on the particularly dangerous assumption in this case that one of the hypotheses tested is the true data-generating distribution. This means that the theory does not take into account the effect of having to fit the hypotheses as models to data, and hence whatever has been deduced must have a fundamental defect. The second fundamental flaw in the theory is that there is no rational quantified way to assess the confidence in the test result arrived. The common test is to decide between the favored *null hypothesis* and either a single opposing hypothesis or, more generally, one of an uncountable number of them, in the case of the so-called composite hypothesis. The null hypothesis would be abandoned only when the data fall in the so-called *critical region*, whose probability under the null hypothesis is less than a certain *level* of the test, such as the commonly used value .05. While clearly there is a considerable confidence in rejecting the null hypothesis when the data fall far in the tails, because the data then are not likely to be typical under the null hypothesis, we have little confidence in accepting it when the data fall outside the critical region. However, what is the point in putting a sharp boundary based only on such obvious and vague considerations? After all, an epsilon variation in the data can swing the decision one way or the other. This single unwarranted act undoubtedly has caused wrong real-life decisions with literally life-and-death implications.

The infamous level of test and its implied confidence is restated in the theory in another essentially equivalent way in terms of the *power* of the test. The power is simply the probability the opposing hypothesis, or one of them in case of a *composite* hypothesis, assigns to the critical region. Since generally the power as a function of the opposing hypothesis grows as the data fall farther and farther in the tails of the null hypothesis, what one can conclude is the same as above: the farther in the tails the data fall, the larger the power and more likely the data will look like it was generated typically by some opposing hypothesis. The theory has other intricate definitions specifying desirable conditions, which invariably can be satisfied only in cases where the conditions are obvious. One example is the *unbiased* test. Just about the only useful result in the theory, as far as I can see, is the celebrated Neyman–Pearson lemma, which states the obvious – that the best critical region S for testing the null hypothesis P_1 against a single opposing hypothesis P_2 is the one where the sum $P_1(\bar{S}) + P_2(S)$ is maximized, where $\bar{S}$ is the complement of S. This is generalized in the obvious way to the case where a level is specified. Also, equivalently, the lemma can be stated in terms of the sum of the two types of errors, which should be minimized, where *type I* error refers to $P_1(S)$ and *type II* to $P_2(\bar{S})$, and called enticingly errors when one or the other hypothesis is true.

We conclude this criticism of the commonly accepted testing theory with a brief comment on the level of significance and the reason it presumably has been introduced. Suppose the test is to find out whether a drug A is good for the common cold. Then, even when the data fall in the tail but outside the critical region, it does not matter too much that we accept a useless drug. However, suppose the drug is intended to cure a serious illness, but it has serious side effects. Now it matters whether we accept or reject the null hypothesis. It seems that we always should find out first the optimal decision boundary by the lemma between the null hypothesis and the opposing ones and the level it gives, and see where the data fall. Then, if we wish to tamper with the decision boundary we can rationally do so only by taking into account the effects of the resulting errors rather than adopting a sort of universal boundary for all cases. It is amazing that a sharp decision boundary, provided by the level of the test, has been accepted by the establishment for so long, without any firm basis to justify the choice.

In this section we discuss a different hypothesis testing procedure, in which hypotheses are models to be fitted to the data. It seems that the real issue in hypothesis testing is to be able to measure how well models fitted to the data are separated. In case of just two models, the problem amounts to calculating the two error probabilities and determining the decision boundary for which the sum of the error probabilities is minimized – i.e., the Neyman–Pearson lemma. The difficult case is when we have a parametric class of models, which is what the *composite* hypothesis should be, one of which is taken as the null hypothesis. The central problem then becomes how to partition the parameter space into at most a countable number of equivalence classes such that any two adjacent models can be *optimally* distinguished from a given amount of data in a measure that is intuitively acceptable and can also be formally justified.

For testing a null hypothesis against the opposing ones, we pick the null hypothesis $f(x^n; \theta^i)$ specified by the center of one of the equivalence classes for the optimal $\hat{d} = 3k$. The opposing composite hypothesis consists of all the other models defined by the centers $\theta^j \neq \theta^i$. If the data fall within $B_{\hat{d}/n}(i)$, we accept the null hypothesis. If they fall within some $B_{\hat{d}/n}(j)$ for $j \neq i$, we reject the null hypothesis, and accept the hypothesis $f(x^n; \theta^j)$.

The confidence in the test can be measured in terms of the two probabilities $P(B_{\hat{d}/n}(i))$ and $1 - P(B_{\hat{d}/n}(i))$, whatever the winning model specified by θ^j is, or if we wish to have a single index, by the ratio

$$\rho(j) = \frac{P(B_{\hat{d}/n}(j))}{1 - P(B_{\hat{d}/n}(i))} .\tag{9.1}$$

We see that the confidence increases rapidly with the increasing distance from the null hypothesis, which appears reasonable; the adjacent models to it are hardest to distinguish.

Since the exact probabilities are not generally available, we calculate the gaussian approximations as follows. With the change of variables $\eta = \sqrt{n}J^{1/2}(\theta^i)(\theta - \theta^i)$, the integral

$$P(B_{d/n}(i)) = \frac{n^{k/2}|J(\theta^i)|^{1/2}}{(2\pi)^{k/2}} \int_{B_{d/n}(i)} e^{(-n/2)\delta' J(\theta^i)\delta} d\delta$$

becomes

$$P(B_{d/n}(i)) = \prod_{j=1}^{k} \frac{2}{\sqrt{2\pi}} \int_0^{\mu_j} e^{-(1/2)\eta^2} d\eta = \prod_{j=1}^{k} 2\phi\left(\sqrt{d/(k\lambda_j)}\right) - 1 ,\tag{9.2}$$

where λ_j is the j'-th eigenvalue of $J(\theta^i)$; μ_j, one-half of the j'-th side length of the rectangle $B_{d/n}(i)$; and $\phi(p)$ is the cumulative density function of the normal density function of mean zero and unit variance.

Testing of two parts of a common model class, such as testing for the arithmetic mean inequality $\hat{m} < m_0$ against $\hat{m} \geq m_0$ in the class of gaussians with a known or unknown variance, is not much different. Take the null hypothesis as the gaussian density function with mean m_0 and either a given or not given variance, say, the former for simplicity. Find the interval for $\hat{d} = 3$ whose center is occupied by m_0. If $\hat{m}$ falls outside this interval and is less than m_0, accept the former hypothesis $\hat{m} < m_0$; else reject it. Hence, in particular, the hypothesis $\hat{m} \geq m_0$ is accepted even if $\hat{m}$ is smaller than m_0 but falls within the interval that includes m_0. The reason is that points within this interval cannot be distinguished from the center.

Example: Bernoulli class

Suppose the null hypothesis is $P(x = 0) = p = i/n$. The width of the equivalence class of the null hypothesis is by (6.31)

$$|B_{3/n}(i)| = \left(\frac{12}{n}\right)^{1/2} ((i/n)(1 - i/n))^{1/2} .\tag{9.3}$$

Its probability is by (9.2)

$$P(B_{3/n}(i)) = 2\phi([3(i/n)(1 - i/n)]^{1/2}) - 1 .$$

For $i/n = 1/3$ the probability of the interval is about 0.58, which appears reasonable. If the data fall within $B_{3/n}(i)$, the confidence index is $P(B_{3/n}(i))/[1 - P(B_{3/n}(i))]$, or in this example about 1.4. If it falls, say, in one of the two adjacent intervals, the confidence is about twice as great.

Example: The class of normal density functions

Consider the case where the null hypothesis is $\mu = 0$ and $\tau = 1$, while the composite hypothesis consists of all other models such that $\mu \neq 0$ and $\epsilon \leq \tau \neq 1$. Putting $\theta = (0, 1)$ we have for $\hat{d} = 6 - 36/n$

$$|B_{\hat{d}/n}(\theta)| = \frac{24\sqrt{2}(1 + 6/n)}{n} . \tag{9.4}$$

There are four nearest optimally distinguishable neighbors, the centers of the four rectangles with the coordinates $2\sqrt{2\hat{d}/n}$ and $-2\sqrt{2\hat{d}/n}$ in the τ direction and $2\sqrt{\hat{d}/n}$ and $-2\sqrt{\hat{d}/n}$ in the μ direction. We take, of course, n so large that $2\sqrt{2\hat{d}/n} < 1 - \epsilon$. There are also four rectangles in the corners of the null hypothesis rectangle $B_{\hat{d}/n}(\theta)$ whose distances from the center $(0, 1)$ are almost as short as the shortest ones.

The null hypothesis is accepted if $(\hat{\mu}(y^n), \hat{\tau}(y^n))$ falls within $B_{\hat{d}/n}(\theta)$, and rejected otherwise. If the null hypothesis is accepted, the confidence index is given by

$$\rho(\theta) = \frac{P(B_{\hat{d}/n}(0, 1))}{1 - P(B_{\hat{d}/n}(0, 1))} ,$$

where the probabilities are calculated with the null hypothesis. If again, $(\hat{\mu}(y^n), \hat{\tau}(y^n))$ falls in, say, $B_{\hat{d}/n}(j)$, then the confidence is given by

$$\rho(j) = \frac{P(B_{\hat{d}/n}(j))}{1 - P(B_{\hat{d}/n}(j))} .$$

Clearly, if this rectangle is not one of the eight neighboring ones, we may have virtual certainty in rejecting the null hypothesis.

9.2 Universal Tree Machine and Variable-Order Markov Chain

The problem in getting the *NML* model for Markov chains and for tree machines, even in the asymptotic form in (6.6), is the integration of the Fisher information needed. The information itself is known [2] –

$$|I(\theta)| = \prod_{s \in S} P_s^{d-1} \prod_{i=0}^{d-1} 1/P(i|s) ,$$

where $\theta = \{P_{i|s}\}$ consists of the conditional probabilities at the states which determine the time-invariant state probabilities P_s by the usual equations.

Due to difficulties in working out the *NML* universal model for Markov chains, we construct a predictive universal model. For this we need to consider predictive encoding of binary strings modeled by Bernoulli processes, whose symbol probability $P(x = 0) = p$ is not known. In other words, the class of processes considered is $\mathcal{B} = \{P(x^n; p)\}$, where

$$P(x^n; p) = p^{n_0}(1 - p)^{n - n_0}$$

and n_0 denotes the number of 0's in the string x^n.

Encode the very first symbol x_1 with the probability $\frac{1}{2}$. Knowing now something about the symbol occurrences, encode the next symbol x_2 with the probability $P(x_2|x_1) = 2/3$, if $x_2 = x_1$, and, of course, with $\frac{1}{3}$, otherwise. In general, then, put

$$P(x_{t+1} = 0|x^t) = \frac{n_0(x^t) + 1}{t + 2} , \tag{9.5}$$

where $n_0(x^t)$ denotes the number of times the symbol 0 occurs in the string x^t. Such a scheme was reputedly invented by Laplace, when he was asked for the probability that the sun would rise tomorrow. It is an easy exercise for the reader to show that this scheme defines the following probability for a sequence x^n:

$$P(x^n) = \frac{n_0!(n - n_0)!}{(n + 1)!} = \binom{n}{n_0}^{-1} (n + 1)^{-1} . \tag{9.6}$$

Hence the ideal code length is

$$L(x^n) = -\log P(x^n) = \log \binom{n}{n_0} + \log(n + 1) . \tag{9.7}$$

Exactly the same probability results from the formula for the marginal distribution

$$P(x^n) = \int_0^1 P(x^n; p)dp = \int_0^1 p^{n_0}(1 - p)^{n - n_0}dp , \tag{9.8}$$

with a uniform prior for the probability p. Such an integral is a Dirichlet's integral.

In particular for strings where n_0/n is close to zero or unity, there is a better estimate for the conditional "next" symbol probability than in (9.5), namely,

$$P(x_{t+1} = 0|x^t) = \frac{n_0(x^t) + 1/2}{t + 1} , \tag{9.9}$$

which is due to Krichevsky and Trofimov. We are now ready to describe an algorithm, called Algorithm Context, introduced in [53] and analyzed in [83], which is universal in a large class of Markov processes. In particular, it provides coding of any string over a given alphabet in such a manner that not only does the per symbol code length approach that of the data-generating Markov process, whatever that process is, but the approach is the fastest possible. We describe first the algorithm for the binary alphabet.

The algorithm has two stages, which can be combined but which we describe separately – namely, Algorithm A for growing a tree, and algorithms for tree pruning, or more accurately, for Choice of the Encoding Node.

Algorithm A

1. Begin with one-node tree, marked with counts $(c_0, c_1) = (1, 1)$ and code length $L(\lambda) = 0$.
2. Read next symbol $x_{t+1} = i$. If none exists, exit. Otherwise, climb the tree by reading the past string backwards $x_t, x_{t-1}, \ldots$ and update the count c_i by unity and the code length $L(s)$ by $-\log P(i|s)$ obtained from (9.5) or (9.9) at every node s met until one of the two following conditions is satisfied:
3. If the node whose count c_i becomes two after the update is an internal node, go to 2. But if it is a leaf, create two new nodes and initialize their counts to $c_0 = c_1 = 1$. Go to 2.

This portion of the algorithm creates a lopsided tree in general, because the tree grows along a path which is traveled frequently. Further, each node s represents a "context", in which symbol i occurs about c_i times in the entire string. In fact, since the path from the root to the node s is a binary string $s = i_1, i_2, \ldots, i_k$, the substring $i_k, \ldots, i_1, i$ occurs in the entire string x^n very close to $c(i|s)$ times, where $c(i|s)$ is the count of i at the node s. Notice that the real occurrence count may be a little larger, because the substring may have occurred before node s was created. Also, the following important condition is satisfied:

$$c(i|s0) + c(i|s1) = c(i|s) , \qquad (9.10)$$

for all nodes whose counts are greater than 1. Finally, the tree created is complete. There exist versions of the algorithm which create incomplete trees.

Choice of encoding nodes

Since each symbol x_{t+1}, even the first, occurs at least in one node s, we can encode it with the ideal code length given by Equation (9.5), where the counts are those at the node s. If a symbol occurs in node s, other than the root, it also occurs in every shorter node on the path to s, which raises the question which node we should pick. There are more than one way to select this "encoding" node, and we describe two.

We use the notation $x^t(s)$ for the substring of the symbols of the "past" string x^t that have occurred at the node s and $L(x^t(s))$ for its code length, which is computed predictively by use of the conditional probabilities (9.5) or (9.9).

Rule 1

For x_{t+1} pick the first node s such that

$$L(x^t(s)) \leq \sum_{i=0}^{1} L(x^t(si)).$$

Notice that both sides refer to the code length for the same symbols – the left-hand side when the symbols occur at the father node, and the right-hand side when they occur at the son nodes. The rule, then, finds the first node where the son nodes' code length is no longer better than the father node's code length. Clearly, because the algorithm does not search the entire path to the leaf, such a strategy may not find the very best node.

If we are willing to find the optimal encoding nodes in two passes through the data, we can find the optimal subtree of any complete tree, where each node is marked with a code length for the symbols that occur at it, and where the symbols that occur at the node also occur at the son nodes. For the final tree, say, $\mathcal{T}$, obtained by Algorithm A, this is insured by Equation (9.10). We write now the code lengths $L(x^t(s))$ more simply as $L(s)$.

Algorithm PRUNE

1. Initialize: For the tree $\mathcal{T}$, put $\bar{S}$ as the set of the leaves. Calculate $I(s) = L(s)$ at the leaves.
2. Recursively, starting at the leaves, compute at each father node s

$$I(s) = \min\{L(s), \sum_{j=0}^{1} I(sj)\} . \tag{9.11}$$

If the first element is smaller than or equal to the second, replace the sons in the set $\bar{S}$ by the father; otherwise, leave $\bar{S}$ unchanged.
3. Continue until the root is reached.

It is easy to convert Algorithm Context into a universal code. All we need is to have an arithmetic coding unit, which receives as the input the "next" symbol x_{t+1} and the predictive probability (9.9) at its encoding node, say, $s^*(t) = s^*(x^t)$, rewritten here

$$P(x_{t+1} = 0|s^*(t)) = \frac{c(0|x^t(s^*(t))) + 1/2}{|x^t(s^*(t))| + 1} . \tag{9.12}$$

Notice that only one coding unit is needed – rather than one for every encoding node. This is because an arithmetic code works for any set of conditional probabilities defining a random process, and, clearly, the conditional probabilities of Algorithm Context define a process by

$$P(x^n) = \prod_{t=0}^{n-1} P(x_{t+1}|s^*(t)) , \qquad (9.13)$$

where $s^*(0) = \lambda$, the root node.

The algorithm generalizes to nonbinary alphabets $\{0, 1, \ldots, d-1\}$ in a straightforward way. The conditional probability rule (9.9) is generalized thus:

$$P(x_{t+1} = i|x^t) = \frac{c_i(x^t) + 1/d}{t+1} . \qquad (9.14)$$

Algorithm B

1. Begin with one-node tree, marked with counts $(c_0, \ldots, c_{d-1}) = (0, \ldots, 0)$ and code length $L(\lambda) = 0$.
2. Read next symbol $x_{t+1} = i$. If none exists, exit. Otherwise, climb the tree by reading the past string backwards $x_t, x_{t-1}, \ldots$ and update the count c_i by unity and the code length $L(s)$ by $-\log P(i|s)$ obtained from (9.14) at every node s met until one of the two following conditions is satisfied:
3. If the node whose count c_i becomes one after the update is an internal node, go to 2. But if it is a leaf, create d new nodes and initialize their counts to $c_i = 0$. Go to 2.

This time when the counts are positive the father node's count of each symbol exceeds the sum of the sons' counts by at most unity

$$\sum_j c(i|sj) + 1 = c(i|s) . \qquad (9.15)$$

The pruning may be done the same way as in the binary case.

What is the code length of this universal code? To answer that question we assume that the data are generated by a tree machine with K leaves. One can show [83] that the mean ideal code length resulting from the universal code defined by a modification of the rule (9.12) satisfies

$$\frac{1}{n} EL_{TM}(X^n) = H(X) + \frac{K}{2n} \log n + o(n^{-1} \log n) ,$$

where $H(X)$ is the entropy of the source. Moreover, by the theorem in Strong Optimality no universal code exists where the mean per symbol code length approaches entropy faster.

9.2.1 Extension to time series

Since the tree grows only at repeated symbol occurrences, the algorithm given will not produce a tree other than depth one for sequences over a large alphabet such as one resulting from quantized real numbers unless the sequences are enormously long. The solution to this problem is to use two quantizations – a coarse precision for the tree growth and the other for encoding the high-precision data at the nodes. We may consider the leaves as providing a partition of the space of sequences, each node defining its own predictor or encoder depending on the application.

Consider a high-precision data string $x^n = x_1, x_2, \ldots, x_n$. Let $x \mapsto \bar{x}$ denote the map when x is quantized to a low precision number, perhaps even binary: $\bar{x} = 1$, if $x > 0$, and $\bar{x} = 0$, if $x \leq 0$. We apply the tree growth algorithm above to the data $\bar{x}^n = \bar{x}_1, \bar{x}_2, \ldots, \bar{x}_n$.

Before considering pruning of the tree, we assign a probability function to the nodes s, for instance, a gaussian density function quantized to the precision in which the original data were given. Hence, we have a conditional probability function $P(x_{t+1}|x^t; \hat{x}_{t+1}, \hat{\tau}_s)$, where $\hat{\tau}_s$ denotes the estimate of the variance obtained from the data points x^t when $\bar{x}_t \bar{x}_{t-1} \ldots = s$, and for instance for AR models, we may write

$$\hat{x}_{t+1} = \hat{a}_{s0} + \hat{a}_{s1}x_t + \hat{a}_{s2}x_{t-1} + \ldots + \hat{a}_{sk}x_{t-k-1} , \tag{9.16}$$

where the coefficients are determined from the occurrences of data at s by the least-squares technique.

Since each node selects data of certain narrow type, it may well happen that the covariance matrix needed to invert to get the coefficients $\hat{a}_{si}$ by minimization is ill conditioned, and a suitable technique such as singular component decomposition is needed to find the minimizing parameters. An interesting phenomenon occurs if the order k is just one symbol larger than the path length to the node s. The effect is like smoothing a piecewise linear partition of the context space.

The pruning of the tree may be done just as described above, or it may done in terms of the minimized quadratic criterion by which the coefficients $\hat{a}_{si}$ are determined. The result is a universal nonlinear AR time series model.

Example. This work was done with Gilbert Furlan. The data $x^n = x_1, \ldots, x_{500}$ were generated by the nonlinear equation

$$x_{t+1} = -.22|x_t|^{1/2}x_{t-1} + .1|x_{t-2}|^{1.3} + e_{t+1} ,$$

where e_t was iid gaussian noise of zero mean and variance $\tau_e = 4.25$. This gave the variance $\tau_x = 6.18$ to the x-process.

We constructed the universal context model above with the course precision as binary, and with a tree of depth varying from two to three, we predicted the data with the linear least-squares predictor (9.16), the coefficients determined from all the data at each leaf. The prediction error turned out to be $\hat{\tau}_x = 5.30$. We also generated a new batch of data y^n of the same length, and with the tree and the coefficients determined from the original data we calculated the prediction error. It came out as $\hat{\tau}_y = 4.85$. Clearly, the new data happened to be easier to predict. In fact, the original system is close to instability, and the sample data vary quite a bit. Finally, we fit an AR model (see subsection AR models below) by the predictive principle, which gave $\hat{k} = 2$ as the optimal order, and the prediction error for the original data as 5.72. We then applied the optimal linear AR model to the new data y^n and got 5.39 as the prediction error. The universal tree model performed clearly better than the optimal linear model.

Example. The so-called logistic map defined by the recursion

$$x_{t+1} = 4x_t(1 - x_{t-1}) + e_{t+1}$$

is an example of a chaotic process, when the data x_t are quantized to any precision, such that e_t may be viewed as the quantization error. The function $4x(1 - x)$ is a concave quadratic function with the maximum 1 at $x = 1/2$, . which vanishes at $x = 0$ and $x = 1$.

We applied Algorithm Context to the binary data $\bar{x}^n$, where $\bar{x}_i = 0$, when $x_i \in (0, 1/2]$, and $\bar{x}_i = 1$, when $x_i \in (1/2, 1]$, and grew the tree only to the balanced tree of depth two. Actually, the tree could be optimized, but it was not done. The four leaves partition the data into four equivalence classes by $\{x^t : x_{t-1}, x_t \mapsto \bar{x}_{t-1}, \bar{x}_t\}$, or in words: all sequences for which the last two symbols get mapped to one of the four corresponding nodes $s = \bar{x}_{t-1}, \bar{x}_t$ are equivalent. To the data $\{x_{st}\}$ at each node or in the corresponding equivalence class fit an AR model of optimal order by

$$\min_{\{a_i\}} \sum_i (x_{st} - a_0 - a_{s1}x_{s(t-1)} - \ldots)^2 .$$

The result is a nonlinear representation of the data

$$x_{t+1} = \hat{a}_{s0} + \hat{a}_{s1}x_{st} + \hat{a}_{s2}x_{s(t-1)} + \hat{a}_{s3}x_{s(t-2)} + \epsilon_{t+1} = \hat{F}(x_t) + u_{t+1} , \quad (9.17)$$

where the index s refers to the node determined by $\bar{x}^t$, and the function $\hat{F}$ is defined by expressing x_{t-1} and x_{t-2} in terms of x_t and putting the terms x_{t-i} for $i > 2$ within u_{t+1}. Quite remarkably, the function $\hat{F}(x)$ is a virtual duplicate of $4x(1 - x)$; it is slightly narrower. One would perhaps expect a peacewise linear approximation of the quadratic function.

Incidentally, one can find just about all the usual parameters in the statistical chaotic theory by a straightforward application of Algorithm Context, which otherwise are determined by various special means. Further details of application of Algorithm Context to chaotic processes can be found in [59].

9.3 Linear Regression

The linear least-squares regression problem, such as polynomial fitting, is a fundamental modeling problem, for which we can give an exact formula for the stochastic complexity and hence make it applicable even for small data sets. Due to the importance of the problem, we discuss it in some detail. This section is an updated version of [61].

We consider the basic linear regression problem, where we have data of type $(y_t, x_{1t}, x_{2t}, \dots)$ for $t = 1, 2, \dots, n$, and we wish to learn how the values x_{it}, $i = 1, 2, \dots, m$, of the *regressor* variables $\{x_i\}$ influence the corresponding values y_t of the *regression* variable y. There may be a large number of the regressor variables, even up to n, and the problem of interest is to find out which subset of them may be regarded to be the most important. This is clearly a very difficult problem, because we must be able to compare the performance of subsets of different sizes. Traditionally it is done by testing hypotheses of the type where some of the regressor variables are removed from the full set, but such testing does not take into account the facts that whatever is tested are models rather than "true" distributions and hence subject to errors, which moreover depend on the complexities of the hypothesized models, which, in turn, depend on the number of regressor variables included.

We fit a linear model of type

$$y_t = \beta' \underline{x}_t + \epsilon_t = \sum_{i \in \gamma} \beta_i x_{it} + \epsilon_t \,, \tag{9.18}$$

where $\gamma = \{i_1, \dots, i_k\}$, $1 \leq k \leq \min\{m, n-1\}$ denotes a subset of the indices of the regressor variables; the prime denotes transposition, and for the computation of the required code lengths the deviations ϵ_t are modeled as samples from an iid Gaussian process of zero mean and variance $\tau = \sigma^2$, also as a parameter. In such a model, the response data $y^n = y_1, \dots, y_n$ are also normally distributed with the density function

$$f(y^n; \gamma, \beta, \tau) = \frac{1}{(2\pi\tau)^{n/2}} e^{-\frac{1}{2\tau} \sum_t (y_t - \beta' \underline{x}_t)^2} \,, \tag{9.19}$$

where $X'_\gamma = \{x_{it} : i \in \gamma\}$ is the $k \times n$ matrix defined by the values of the regressor variables with indices in γ. Write $Z_\gamma = X'_\gamma X_\gamma = n\Sigma_\gamma$, which is taken to be positive definite. The development for a while will be for a fixed γ, and we drop the subindex in the matrices above as well as in the parameters. The maximum-likelihood solution of the parameters is given by

$$\hat{\beta}(y^n) = Z^{-1} X' y^n \tag{9.20}$$

$$\hat{\tau}(y^n) = \frac{1}{n} \sum_t (y_t - \hat{\beta}'(y^n) \underline{x}_t)^2 \,. \tag{9.21}$$

The density function (9.19) admits an important factorization as follows:

$$f(y^n; \gamma, \beta, \tau) = f(y^n|\hat{\beta}, \hat{\tau})p_1(\hat{\beta}; \beta, \tau)p_2(n\hat{\tau}/\tau; \tau)\frac{n}{\tau} \tag{9.22}$$

$$f(y^n|\gamma, \hat{\beta}, \hat{\tau}) = (2\pi)^{\frac{k-n}{2}} n^{-n/2}|\Sigma|^{-1/2}\Gamma\left(\frac{n-k}{2}\right) 2^{\frac{n-k}{2}}\hat{\tau}^{1-\frac{n-k}{2}} \tag{9.23}$$

$$p_1(\hat{\beta}; \beta, \tau) = \frac{n^{k/2}|\Sigma|^{1/2}}{(2\pi\tau)^{k/2}}e^{-\frac{n}{2\tau}(\hat{\beta}-\beta)'\Sigma(\hat{\beta}-\beta)} \tag{9.24}$$

$$p_2(n\hat{\tau}/\tau; \tau) = (n\hat{\tau}/\tau)^{\frac{n-k}{2}-1}e^{-n\frac{\hat{\tau}}{2\tau}}2^{-\frac{n-k}{2}}\Gamma^{-1}\left(\frac{n-k}{2}\right). \tag{9.25}$$

With $\theta = (\beta, \tau)$ we can write $f(y^n; \gamma, \beta, \tau) = f(y^n; \gamma, \theta)$, and since $f(y^n; \gamma, \theta) = f(y^n, \hat{\theta}(y^n); \gamma, \theta)$, we can write it as the product of the marginal density of $\hat{\theta}$

$$p(\hat{\theta}; \gamma, \theta) = p_1(\hat{\theta}; \theta)p_2(n\hat{\tau}/\tau; \tau)\frac{n}{\tau}$$

and the conditional density of y^n given $\hat{\theta}$

$$f(y^n; \gamma, \theta) = f(y^n|\hat{\theta}(y^n); \gamma, \theta)p(\hat{\theta}(y^n); \gamma, \theta). \tag{9.26}$$

The factor p_1 is normal with mean β and covariance $\frac{\tau}{n}\Sigma^{-1}$, while p_2 is the χ^2 distribution for $n\hat{\tau}/\tau$ with $n - k$ degrees of freedom. Moreover, they are independent, and the statistic $\hat{\theta}$ is said to be sufficient. This is because $f(y^n|\hat{\theta}(y^n); \gamma) = h(y^n)$ depends only on the data and not on θ.

The two cases $k = 0$ and $0 < k \leq \min\{m, n-1\}$ require somewhat different treatments, and we consider first the latter case.

The *NML* density function is given by

$$\hat{f}(y^n; \gamma) = \frac{f(y^n; \gamma, \hat{\beta}(y^n), \hat{\tau}(y^n))}{\int_{Y(\tau_0, R)} f(z^n; \gamma, \hat{\beta}(z^n), \hat{\tau}(z^n))dz^n}, \tag{9.27}$$

where y^n is restricted to the set

$$Y(\tau_0, R) = \{z^n : \hat{\tau}(z^n) \geq \tau_0, \hat{\beta}'(y^n)\Sigma\hat{\beta}(y^n) \leq R\}. \tag{9.28}$$

Further, $Y(\tau_0, R)$ is to include y^n.

The numerator in Equation (9.27) has a very simple form –

$$f(y^n; \gamma, \hat{\beta}(y^n), \hat{\tau}(y^n)) = 1/(2\pi e\hat{\tau}(y^n))^{n/2}, \tag{9.29}$$

and the problem is to evaluate the integral in the denominator.

Integrating the conditional $f(y^n|\hat{\theta}(y^n); \gamma, \theta) = h(y^n)$ over y^n such that $\hat{\theta}(y^n)$ equals any fixed value $\hat{\theta}$ yields unity. Therefore with $p(\hat{\theta}; \gamma, \hat{\theta}) \equiv g(\hat{\tau})$ we get from the expression for the χ^2 density function in (9.22),

$$C(\tau_0, R) = \int_{Y(\tau_0, R)} f(y^n; \hat{\theta}(y^n)) dy^n \tag{9.30}$$

$$= A_{n,k} \int_{\tau_0}^{\infty} \hat{\tau}^{-\frac{k+2}{2}} d\hat{\tau} \int_{B_R} d\beta \tag{9.31}$$

$$= A_{n,k} V_k \frac{2}{k} \left(\frac{R}{\tau_0}\right)^{k/2}, \tag{9.32}$$

where $B_R = \{\beta : \beta' \Sigma \beta \le R\}$ is an ellipsoid,

$$V_k R^{k/2} = |\Sigma|^{-1/2} \frac{2\pi^{k/2} R^{k/2}}{k\Gamma(k/2)}, \tag{9.33}$$

its volume, and

$$A_{n,k} = \frac{|\Sigma|^{1/2}}{\pi^{k/2}} \frac{\left(\frac{n}{2e}\right)^{\frac{n}{2}}}{\Gamma\left(\frac{n-k}{2}\right)}. \tag{9.34}$$

We then have the *NML* density function itself for $0 < k \le m$ and $y = y^n$:

$$-\log \hat{f}(y; \gamma, \tau_0, R) = \frac{n}{2} \ln \hat{\tau} + \frac{k}{2} \ln \frac{R}{\tau_0} - \ln \Gamma\left(\frac{n-k}{2}\right)$$

$$- \ln \Gamma\left(\frac{k}{2}\right) + \ln \frac{4}{k^2} + \frac{n}{2} \ln(n\pi). \tag{9.35}$$

We wish to get rid of the two parameters R and τ_0, which clearly affect the criterion in an essential manner, or rather we replace them with other parameters which influence the relevant criterion only indirectly. We can set the two parameters to the values that minimize (9.35): $R = \hat{R}$, and $\tau_0 = \hat{\tau}$, where $\hat{R} = \hat{\beta}'(y)\Sigma\hat{\beta}(y)$. However, the resulting $\hat{f}(y; \gamma, \hat{\tau}(y), \hat{R}(y))$ is not a density function. We can, of course, correct this by multiplying it by a prior $w(\hat{\tau}(y), \hat{R}(y))$, but the result will be a density function on the triplet $y, \hat{\tau}(y), \hat{R}(y)$, which is not quite right. We do it instead by the same normalization process as above:

$$\hat{f}(y; \gamma) = \frac{\hat{f}(y; \gamma, \hat{\tau}(y), \hat{R}(y))}{\int_Y \hat{f}(z; \gamma, \hat{\tau}(z), \hat{R}(z)) dz}, \tag{9.36}$$

where the range Y will be defined presently. By (9.26) and the subsequent equations we also have the factorization

$$\hat{f}(y; \gamma, \tau_0, R) = f(y|\gamma, \hat{\beta}, \hat{\tau}) g(\hat{\tau}) / C(\tau_0, R)$$

$$= f(y|\gamma, \hat{\beta}, \hat{\tau}) \frac{k}{2} \hat{\tau}^{-k/2-1} V_k^{-1} \left(\frac{\tau_0}{R}\right)^{k/2}. \tag{9.37}$$

As above we can now integrate the conditional while keeping $\hat{\beta}$ and $\hat{\tau}$ constant, which gives unity. Then by setting $\tau_0 = \hat{\tau}$ and $R = \hat{R}$, we integrate the resulting function of $\hat{\tau}$ over a range $[\tau_1, \tau_2]$ and $\hat{R}(\beta)$ over the volume between the hyper-ellipsoids bounded by $[R_1 \leq \hat{R} \leq R_2]$. All told we get

$$\int_Y \hat{f}(z; \gamma, \hat{\tau}(z), \hat{R}(z)) dz = \frac{k}{2} V_k^{-1} \int_{\tau_1}^{\tau_2} \int_{R_1}^{R_2} \tau^{-1} R^{-k/2} d\tau dV$$

$$= \left(\frac{k}{2}\right)^2 \ln \frac{\tau_2}{\tau_1} \ln \frac{R_2}{R_1} ,$$

where we expressed the volume element as

$$dV = \frac{k}{2} V_k R^{\frac{k}{2}-1} dR .$$

The negative logarithm of $\hat{f}(y; \gamma)$, which we called the stochastic complexity, is then given by

$$- \ln \hat{f}(y; \gamma) = \frac{n-k}{2} \ln \hat{\tau}_\gamma + \frac{k}{2} \ln \hat{R}_\gamma - \ln \Gamma \left(\frac{n-k}{2}\right) - \ln \Gamma \left(\frac{k}{2}\right)$$

$$+ \frac{n}{2} \ln(n\pi) + \ln[\ln \left(\frac{\tau_2}{\tau_1}\right) \ln \frac{R_2}{R_1}] , \qquad (9.38)$$

where we indicate by the subindex γ that we have selected the submatrix of X' consisting of the rows indexed by γ. We mention that a similar criterion to optimize over γ was obtained in [27] with special priors for the hyper-parameters.

The hyper-parameters are to be selected so that the range they define includes the maximum-likelihood estimates, and for such data the term defined by the hyper-parameters can be ignored. Although their effect is small, only like ln ln, the effect is insidious. Since virtually for all data both $\hat{R}(y)$ and $\hat{\tau}(y)$ will be positive, so that we may squeeze positive lower bounds below these values and make the code length valid, the Fisher information matrix could be close to singular. This implies a very slow convergence with unreliable estimates. This happens in particular in the denoising problem, where m is large, and for which bad results have been reported in the case with zero-mean gaussian white noise data. A more important difficulty arises when we wish to compare the stochastic complexity for positive values of k with that for $k = 0$, which we calculate next.

The maximized-likelihood is the same as before, and in the decomposition (9.22) the factor p_1 may be taken as unity. By the same technique as above we get

$$p(\hat{\tau}; \gamma, \hat{\tau}) = \ln \left(\frac{n}{2e}\right)^{n/2} \Gamma^{-1} \left(\frac{n}{2}\right) \frac{n}{\hat{\tau}} ,$$

and its integral from τ_3 to τ_4

$$C_{0,n} = \ln \left(\frac{n}{2e}\right)^{n/2} \Gamma^{-1} \left(\frac{n}{2}\right) \ln \left(\frac{\tau_4}{\tau_3}\right) . \qquad (9.39)$$

Then the stochastic complexity is given by

$$- \ln \hat{f}(y; 0) = \frac{n}{2} \ln n\pi\hat{\tau} - \ln \Gamma \left(\frac{n}{2} \right) + \ln \ln \frac{\tau_4}{\tau_3} . \qquad (9.40)$$

Since $n\hat{\tau} = y'y = \sum_t y_t^2$, we see in (9.40) that the hyper-parameters $n\tau_1$ and $n\tau_2$ define the range for the sum of the squares of the data sequence. Because $y'y = n(\hat{R} + \hat{\tau})$ we see that if we apply the two criteria (9.38) and (9.40) to the same data, we need to have $y'y/n = R_1 + \tau_1 \geq \tau_3$ and $y'y/n = R_2 + \tau_2 \leq \tau_4$. Hence, put $\tau_3 = \tau_1 + R_1$ and $\tau_4 = \tau_2 + R_2$. To simplify the notations we pick $R_1 = \tau_1 = a$, $R_2 = \tau_2 = b$, and $b > a$.

By applying Stirling's approximation to the Γ-functions in (9.38) and adding the code length for γ (2.9),

$$L(\gamma) = \min\{m, [\ln \binom{m}{k} + \ln k + \log \ln(em)]\} , \qquad (9.41)$$

we get

$$- \ln \hat{f}(y; \gamma) = \frac{n-k}{2} \ln \frac{\hat{\tau}}{n-k} + \frac{k}{2} \ln \frac{\hat{R}}{k} + \frac{1}{2} \ln(k(n-k))$$
$$+ 2 \ln \ln \frac{b}{a} + L(\gamma) + \frac{n}{2} \ln(2n\pi e) - 3 \ln 2 , \qquad (9.42)$$

where we ignored the small difference between the Γ function and its approximation. For many of the usual regression problems the code length for γ is small and can be omitted.

Similarly (9.40) is given by

$$- \ln \hat{f}(y; 0) = \frac{n}{2} \ln(2\pi e \hat{\tau}) - \ln \frac{n}{2^{3/2}\sqrt{\pi}} + \ln \ln \frac{b}{a} . \qquad (9.43)$$

We see that in the comparison of the two criteria the term $\ln \ln \frac{b}{a}$ still remains. By the theorem below for each γ (see also the explanation in subsection on orthonormal regression matrices) we should take $\max_\beta \hat{R}_\gamma$, whose smallest value is obtained for $k = 1$ and the largest value for $k = m$, and pick a and b as these extremes. Again, unless these values are extremely small and large, respectively, the term $\ln \ln \frac{b}{a}$ has little effect to the comparison.

There is another useful form of the criterion, obtained from (9.42) by dropping the terms not dependent on k, which was derived by E. Liski [45]. Define

$$n\hat{R} = \sum_{t=1}^{n} y_t^2 - RSS \qquad (9.44)$$

$$RSS = \sum_{t=1}^{n} y_t^2 - n\hat{R} \qquad (9.45)$$

$$S^2 = \frac{RSS}{n-k} \tag{9.46}$$

$$F = \frac{\hat{R}}{kS^2} , \tag{9.47}$$

where S^2 is the unbiased estimate of the variance $\hat{\tau}$, and F is the statistic to test the hypothesis that $\beta = 0$. Then the criterion takes the form

$$\min_{\gamma} \left\{ \frac{n}{2} \ln S^2 + \frac{k}{2} \ln F + \ln(k(n-k)) + 2\ln\ln\frac{b}{a} + L(\gamma) \right\} . \tag{9.48}$$

This shows the crucial adaptation property of the *NML* criterion that the second term, the code length for the model, gets smaller for data where S^2 is large, and, conversely. Experiments show [45] that indeed the penalty for the number of parameters is $O(k\ln n)$ when the data are generated by a model with a small number of parameters, and that the criterion behaves like *AIC*, where the penalty is just the number of parameters, when the data are generated by a model with a large number of parameters. This phenomenon was also noticed in [27].

E. Liski also proved a coordinate free statement of an earlier theorem [61] (proved below in the subsection on orthonormal regression matrices), which we restate in a slightly different form, as follows:

Theorem 21 *For each k the structure $\hat{\gamma}$ that minimizes the NML criterion (9.48) is obtained either by maximizing or minimizing RSS.*

Example: Polynomial fitting

The criterion (9.42) actually incorporates the integral of the square root of Fisher information, see (5.40), which is very important for small amounts of data. We illustrate this with the polynomial fitting problem. (The example was done by Ciprian Giurcaneanu, Technical University of Tampere.)

A signal is generated by a third-order polynomial $y = x^3 - 0.5x^2 - 5x - 1.5$, where the x-values are generated by a uniform distribution in $[-3, 3]$. A gaussian 0-mean noise is added to y-values. The variance is selected to have SNR=10 dB. We see in Table 9.1 the superiority of the *NML* criterion.

9.3.1 Orthonormal regression matrix

A great simplification results when the $m \times n$ regressor matrix X' is orthonormal, so that Σ is the $m \times m$ identity matrix. In fact, convert the rows of X' into an orthonormal set W' by rotation and normalization of the coordinates thus:

$$W' = \Lambda^{-1/2} Q' X' ,$$

where $Q' = Q^{-1}$ and Λ is a diagonal matrix such that $\Lambda^{-1/2} Q' \Sigma Q \Lambda^{-1/2} = I$, the $m \times m$ identity matrix. In the new coordinate system, the ML estimates get transformed into new ones as

Table 9.1. Order estimation of the polynomial model in the example*

Order	Criterion	Sample size								
		25	30	40	50	60	70	80	90	100
$\hat{k} < k$	NML	0	0	0	0	0	0	0	0	0
	BIC	0	0	0	0	0	0	0	0	0
	KICC	0	0	0	0	0	0	0	0	0
$\hat{k} = k$	NML	940	956	963	972	968	977	971	982	980
	BIC	789	844	904	909	925	944	932	957	951
	KICC	933	923	923	901	896	905	895	909	888
$\hat{k} > k$	NML	60	44	37	28	32	23	29	18	20
	BIC	211	156	96	91	75	56	68	43	49
	KICC	67	77	77	99	104	95	105	91	112

*For 1,000 runs the counts show the number of times the order was underestimated $(0 \le \hat{k} \le 2)$, correctly estimated $(\hat{k} = 3)$, and overestimated $(4 \le \hat{k} \le 10)$ by each criterion. (*BIC* is the original crude *MDL* criterion; *KICC* is bias-corrected Kullback information criterion)

$$\hat{\eta} = \Lambda^{-1/2}Q'\hat{\beta} = W'y \, ,$$

We also have

$$\hat{R}_\gamma = \hat{\eta}'\hat{\eta} = \hat{\beta}'\Sigma\hat{\beta} \tag{9.49}$$

$$n\hat{\tau}_\gamma = y'y - \hat{R}_\gamma \, . \tag{9.50}$$

Further, the matrix W may be regarded as a transform between the coefficients $\hat{\eta}$, written now more briefly as $c = c_1, \ldots, c_m$, and the projection $\hat{y}$ of the data y onto the space spanned by the rows of W',

$$\hat{y} = Wc \tag{9.51}$$

$$c = W'\hat{y} = W'y \, . \tag{9.52}$$

We also see that we only need to calculate c for the full matrix W', since for any submatrix W'_γ obtained by the rows indexed by γ, the ML estimates $\hat{\eta} = \hat{c}$ are just the corresponding components of c.

The criterion (9.42) for finding the best subset γ, including the number k of its elements, is then given by

$$\min_\gamma \ln 1/\hat{f}(y; \gamma) = \min_\gamma \left\{ \frac{n-k}{2} \ln \frac{c'c - \hat{S}_\gamma}{n-k} + \frac{k}{2} \ln \frac{\hat{S}_\gamma}{k} \right.$$
$$\left. + \frac{1}{2} \ln(k(n-k)) + 2\ln\ln\frac{b}{a} + L(\gamma) \right\} , \tag{9.53}$$

where $L(\gamma)$ is given in (9.41) and

$$\hat{S}_\gamma = \hat{c}'\hat{c} = \sum_{i\in\gamma} c_i^2 . \tag{9.54}$$

It may seem that the minimization of (9.53) requires a search through all the 2^m subsets of the m basis vectors. However, we have the following theorem [61], which shows that minimization requires only $2m$ evaluations of the criterion

Theorem 22 *For each k, the index set $\hat{\gamma}$ that minimizes (9.53) is given either by the indices $\hat{\gamma} = \{(1), \dots, (k)\}$ of the k largest coefficients in absolute value or the indices $\hat{\gamma} = \{(m - k + 1), \dots, (m)\}$ of the k smallest ones.*

Proof. Let γ be an arbitrary collection of a fixed number of indices k, and let $\hat{S}_\gamma$ be the corresponding sum of the squared coefficients. Let $u_i = c_i^2$ be a term in $\hat{S}_\gamma$. The derivative of $\ln 1/\hat{f}(y; \gamma)$ with respect to u_i is then

$$\frac{d \ln 1/\hat{f}(y; \gamma)}{du_i} = \frac{k}{\hat{S}_\gamma} - \frac{n - k}{c'c - \hat{S}_\gamma} , \tag{9.55}$$

which is nonpositive when $\hat{S}_\gamma/k \geq \hat{T}_\gamma/(n - k)$, where $\hat{T}_\gamma = c'c - \hat{S}_\gamma$, and positive otherwise. The second derivative is always negative, which means that $\ln 1/\hat{f}(y; \gamma)$ as a function of u_i is concave.

If for some γ, $\hat{S}_\gamma/k > \hat{T}_\gamma/(n - k)$, we can reduce $\ln 1/\hat{f}(y; \gamma)$ by replacing, say, the smallest square c_i^2 in γ by a larger square outside of γ, and get another γ for which $\hat{S}_\gamma$ is larger and $\hat{T}_\gamma$ smaller. This process is possible until $\gamma = \{(1), \dots, (k)\}$ consists of the indices of the k largest squared coefficients. Similarly, if for some γ, $\hat{S}_\gamma/k < \hat{T}_\gamma/(n - k)$, we can reduce $\ln 1/\hat{f}(y; \gamma)$ until γ consists of the indices of the k smallest squared coefficients. Finally, if for some γ, $\hat{S}_\gamma/k = \hat{T}_\gamma/(n - k)$, then all the squared coefficients must be equal, and the claim holds trivially.

It seems weird that the criterion would be minimized by the smallest coefficients. However, we have made no assumptions about the data, and practically always in regression problems the smallest coefficients represent noise while the largest represent the "smooth" projection $\hat{y}$. Hence for such data the largest coefficients will be optimal.

9.4 MDL Denoising

Denoising is an important problem with many applications. Specifically, the problem is to remove noise from a data sequence $x^n = x_1, \dots, x_n$ and smooth the data. Traditionally, especially in engineering applications, "noise" is defined as the high-frequency part in the signal, which can be removed by passing the signal through a suitably designed low-pass filter. There is of course the problem of deciding how high the frequency must be to qualify as noise. However, when the technique is applied to a wide variety of data, especially when the smoothed data is expressed in wavelet basis vectors, which are much more flexible than the sinusoidals in traditional Fourier transforms, the idea of uniform frequency over the data is no longer applicable, and the nature of noise gets blurred. One approach taken and pioneered in [17] is to assume that there is a "true" smooth signal, $\bar{x}^n = \bar{x}_1, \ldots, \bar{x}_n$, representing the real information bearing data,

$$x_t = \bar{x}_t + \epsilon_t, \quad t = 1, \ldots, n, \tag{9.56}$$

to which independent normally distributed noise ϵ_t of 0-mean and variance τ is added. The task is to estimate the "true" signal $\bar{x}_t$, expressed in an orthonormal basis $\{\underline{w}'_i = w_{i1}, \ldots w_{in}\}$ for $i = 1, \ldots, n$, and the estimation is done by minimization of the risk criterion $R = \sum_t E(x_t - \bar{x}_t)^2 = n\tau$, where the expectation with respect to the assumed normal distribution is to be estimated from the data.

The first difficulty with this type of approach is the assumption of a "true" signal $\bar{x}$, which in general does not exist at all. For instance, a digitized two-dimensional image of a three-dimensional object, such as of an internal organ, obtained with some imaging device is very difficult to claim to represent a "true" image. Similarly a photograph of a collection of stars with all the interference is all we get, and again it is difficult to claim that any different "true" picture exists. A more serious objection is the estimation of τ. In fact, the regression matrix $W' = \{w_{ij}\}$, defined by the rows $\underline{w}'_i$, has the transpose W as the inverse, and it defines the 1–1 transformation

$$x = Wc$$
$$c = W'x, \tag{9.57}$$

where x and c denote the column vectors of the strings of the data $x' = x_1, \ldots, x_n$ and the coefficients $c' = c_1, \ldots, c_n$, respectively. Because of orthonormality Parseval's equality $c'c = \sum_t c_t^2 = x'x = \sum_t x_t^2$ holds. Accordingly, taking $\bar{x} = x$ gives the natural estimate $n\hat{\tau} = (x - \bar{x})'(x - \bar{x}) = 0$, which is a meaningless minimum for the risk. Hence, the estimation must be done by some arbitrary scheme. The trouble is that the noise gets determined by the estimate, and since it is supposed to represent the noise variance it is needed before the noise is obtained.

We describe in this chapter a different approach to the denoising problem based on the representation of the smooth signal $\hat{x}^n$ as being in one-to-one correspondence with the optimal model in the structure function (Figure 6.2) and hence the noise as the remaining part $x^n - \hat{x}^n$ in the data. Notice that the structure function does not give the noise part as a signal; it just gives the optimal amount of noise. Hence, noise can be defined as that part in the data that has no learnable useful information, which appears to conform with intuition. We discuss two different constructs, one as a special case of the linear quadratic regression problem [61], and the other such that both the noise and the denoised signal, still expressed in terms of wavelets, are modeled by histograms [36]. The latter version should perform especially well when the noise part tends to have a nongaussian distribution, which is the case in certain type of applications (see [36]).

9.4.1 Linear-quadratic denoising

Instead of (9.56) we have the decomposition

$$x_t = \hat{x}_t + e_t, \quad t = 1, \ldots, n , \tag{9.58}$$

where $\hat{x}$ is defined by the optimal model in (9.53), for $m = n$ and the optimizing γ with k components, $0 < k < n$. Instead of transforming a generally huge regressor matrix into an orthonormal one, it is customary to apply a wavelet transform to the data, which avoids even writing down the regressor matrix.

Although by the theorem in the previous subsection the optimal γ consists either of a number of the largest or the smallest coefficients in absolute value, the data in the denoising problem are typically such that the "smooth" denoised curve is simple, and we should find the k largest coefficients $c_{(1)}, \ldots, c_{(k)}$ in absolute value to minimize

$$\min_k \ln 1/\hat{f}(x; \gamma) = \min_k \left\{ \frac{n-k}{2} \ln \frac{c'c - \hat{S}_{(k)}}{n-k} + \frac{k}{2} \ln \frac{\hat{S}_{(k)}}{k} + \frac{1}{2} \ln(k(n-k)) \right.$$

$$\left. + \ln \left[\left(\ln \frac{\tau_2}{\tau_1} \right) \left(\ln \frac{R_2}{R_1} \right) \right] + L(\gamma) \right\} , \tag{9.59}$$

where $\hat{S}_{(k)}$ is the sum of the largest squares of k coefficients, and $L(\gamma)$ is given in (9.41); we also included the hyper-parameters, which may be needed if the criterion is applied to unusual data. This time they restrict the data such that $c_{(1)}^2 \geq R_1$, $c'c - c_{(n)}^2 \leq R_2$, $c_{(n)}^2 \geq \tau_1$, and $c'c - c_{(1)}^2 \leq \tau_2$. These hold if we require $x'x < R_2 = \tau_2$ and $c_{(n)}^2 > R_1 = \tau_1$. The latter condition puts in general the tougher restriction, because the smallest squared coefficient can be small indeed. Since we do not wish to apply the criterion to pure noise data without any other smooth denoised curve than $\hat{x} = 0$, we can ignore the term defined by the hyper-parameters. However, to guard us against poor behavior in extreme data, we simply restrict $\hat{k}$ to the range $[1, \min\{n-1, \alpha n\}]$ for α something like 0.95. For a different derivation of the stochastic complexity criterion with a so-called soft threshold we refer to [67]. Another type of application to images is in [28].

With $\hat{c}^n$ denoting the column vector defined by the coefficients $\hat{c}_1, \ldots, \hat{c}_n$, where $\hat{c}_i = c_i$ for $i \in \{(1), \ldots, (\hat{k})\}$ and zero, otherwise, the signal recovered is given by $\hat{x}^n = W\hat{c}^n$.

We calculate two examples using wavelets defined by Daubechies' N = 6 scaling function.

Example 1

The first example is a case where the *MDL* threshold is close to those obtained with traditional techniques, as well as to the threshold obtained with a rather complicated cross-validation technique. The mean signal $\bar{x}_i$ consists of 512 equally spaced samples of the following function defined by three piecewise polynomials:

$$x(t) = \begin{cases} 4t^2(3 - 4t) & \text{for } t \in [0, .5] \\ \frac{4}{3}t(4t^2 - 10t + 7) - \frac{3}{2} & \text{for } t \in [.5, .75] \\ \frac{16}{3}t(t - 1)^2 & \text{for } t \in [.75, 1] \end{cases}$$

To the data points $\bar{x}_i$ were added pseudorandom normal 0-mean noise with standard deviation of 0.1, which defined the data sequence x_i.

The threshold obtained with the *NML* criterion is $\lambda = 0.246$. This is between the two thresholds called VisuShrink $\lambda = 0.35$ and GlobalSure $\lambda = 0.14$, both of the type derived by Donoho and Johnstone. It is also close to the threshold $\lambda = 0.20$, obtained with the much more complex cross-validation procedure by Nason.

Example 2

In this example the data sequence consists of 128 samples from a voiced portion of speech. The *NML* criterion retains 41 coefficients exceeding the threshold $\lambda = 7.4$ in absolute value. Figure 9.1 shows the original signal together with the information bearing signal extracted by the *NML* criterion. We see that the *NML* criterion has not removed the sharp peaks in the large pulses despite the fact that they have locally high frequency content. They simply can be compressed by use of the retained coefficients, and by the general principle behind the criterion they are not regarded as noise.

9.4.2 Histogram denoising

The actual algorithm, which breaks up the data into several resolution layers in the wavelet transform, is rather complicated, and we contend ourselves to describing only the general idea.

Code length of data with histograms

We need the code length for a sequence $y^n = y_1, y_2, \ldots, y_n$ of real valued data points y_t, quantized to a common precision δ and modelled by a histogram. The data will actually consist of the coefficients obtained by the wavelet transform. Let all y_t fall in the interval $[a, b]$, which is partitioned into m equal-width bins, the width given by $w = R/m$, where $R = b - a$. Let n_i data points fall in the i'th bin. Then the code length of the data string, actually the stochastic complexity, relative to such histograms, is given by

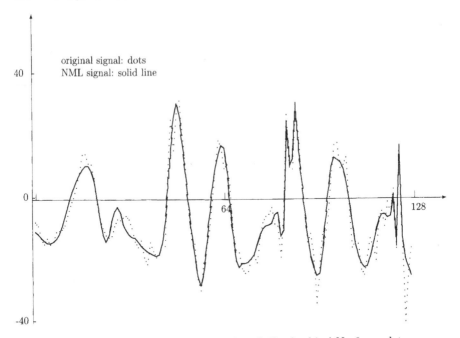

Fig. 9.1. Speech signal smoothed with Daubechies' N=6 wavelet

$$L(y^n|w, m, \delta) = \log\binom{n}{n_1, ..., n_m} + \log\binom{n+m}{n} + n\log(w/\delta), \qquad (9.60)$$

where the logarithms are to base 2 see ([25, 64]). The first term is the code length for encoding the sequence of the m bin indices corresponding to the bins of $y_1, y_2, \ldots, y_n$. The length of this sequence is n. The second term is the code length for the integers n_i. To see this, notice that the positive indices can be encoded as a binary string, which starts with n_1 0's and a 1, followed by n_2 0's and a 1, and so on. The string has m 1's and its length is $n + m$. If we sort all such strings we can encode the index of any one with code length given by the logarithm of their number $\binom{n+m}{n}$. The third term gives the code length for the numbers y_t, quantized to the precision δ. Indeed, each quantized number is one of the w/δ points in its bin. If we add the code length, about $L^*(m)$ bits, (2.4), for the number m we can minimize the conditional code length

$$\min_m \{L(y^n|w, m, \delta) + L^*(m)\}$$

to remove the dependence on m, which leaves only the two numbers w and δ on which the code length is conditioned.

Now we need to consider the code length of the modified data string $\hat{y}^n$, obtained by retaining the coefficients in a subset S of the bins while setting the rest to zero: i.e., the points in the set of the remaining bins $\bar{S}$ are 0. Denote the indices of the bins of the data points $\hat{y}_t$ by $(0), (1), \ldots, (|S|)$, where $|S|$

denotes the number of the bins in the set S. The bin with index (0) is added to contain all the $n_{(0)} = n - k$ 0-points, where k denotes the number of the retained points falling in the bins of S. For instance, if the first retained bin is the fifth bin out of the m bins, then $(1) = 5$. The code length for the sequence of the bin indices is now

$$\log \binom{n}{n_{(1)}, \ldots, n_{(|S|)}, n - k},$$

(9.61)

where $n_{(j)}$ denotes the number of points falling in the bin having the index (j). Then the code length for the string $\hat{y}^n$, given w, m, and δ, is

$$L(\hat{y}^n | m, w, \delta) = \log \binom{n}{n_{(1)}, \ldots, n_{(|S|)}, n - k}$$

$$+ \log \binom{n + |S| + 1}{n} + k \log(w/\delta) + m .$$

(9.62)

The last term is the code length for S, which is one of the 2^m subsets of the bins.

The *MDL*-histo algorithm

We now describe the main denoising algorithm. Let $c^n = c_1, \ldots, c_n$ denote the sequence of coefficients obtained by applying a wavelet transform to the data sequence x^n to be denoised, and let R denote their range. An equal-bin-width histogram H with m bins of width $w = R/m$ is fitted to the coefficients. The number of bins determines the amount of computations needed and it is not optimized. In [36] it is selected to be $m = 7$.

We start the selection process for the retained coefficients. Let S be one of the 2^m subsets of the m bins having k coefficients from the sequence c^n falling in the bins of S. Put $m_1 = |S|$. Define $\hat{c}^n$ as the string of the points falling in the bins of S, and zero otherwise. The code length for $\hat{c}^n$ is given as in (9.62) by

$$L(\hat{c}^n | S, w, \delta) = \log \binom{n}{n_{(1)}, \ldots, n_{(m_1)}, n - k} + \log \binom{n + m_1 + 1}{n} + k \log(w/\delta),$$

(9.63)

where $n_{(j)}$ denotes the number of points falling in the bin of H having the index (j). To clarify these indices, let b_k denote the k-th bin of H for $k = 1, 2, \ldots, 7$, and let the set S of the retained nonzero bins be $S = \{b_2, b_3, b_5, b_7\}$. Then $(1) = 2$, $(2) = 3$, $(3) = 5$, and $(4) = 7$. We also omit the fixed code length $m = 7$ for the set S.

Next we calculate the code length of the residual $\bar{e}^n$. Since the wavelet transform can be inverted, the code length is obtained from the coefficients c^n, in which the coefficients retained so far are set to zero. Let $\bar{c}^n = c^n - \hat{c}^n$

denote the string of the residuals. In this sequence there are $n - k$ nonzero coefficients, provided that all the n coefficients in c^n are nonzero. We model this sequence by an equal-bin-width histogram of M bins. Hence, the common bin width is given by $w = R/M$. Since k coefficients in $\bar{c}^n$ were set to zero, the sequence $\bar{c}^n$, given S, w, δ, and M, can be encoded with the code length

$$L(\bar{c}^n|S, w, \delta, M) = \log \binom{n-k}{\nu_1, \ldots, \nu_M} + \log \binom{n-k+M}{M} + (n-k)\log(w/\delta),$$

(9.64)

where ν_j is the number of points falling in the j'-th bin.

Ignoring the code length for the range R, the code length for the data is given by

$$L(x^n|S, w) = L(\hat{c}^n|S, w, \delta) + L(\bar{c}^n|S, w, \delta, M) + L^*(M) \, .$$

In this the precision δ does not appear at all, and the term $n \log R$ does not depend on S. The subset S is then determined by the following minimization criterion:

$$\min_{S,M} \left\{ \log \binom{n}{n_{(1)}, \ldots, n_{(m_1)}, n-k} + \log \binom{n+m_1+1}{n} \right.$$
$$+ \log \binom{n-k}{\nu_1, \ldots, \nu_M} + \log \binom{n-k+M}{M}$$
$$\left. - n \log M + k_1 \log(MR/(mR_1)) + L^*(M) \right\} \, .$$

(9.65)

There are a number of ways to improve this scheme. One is to employ variable-bin-width histograms, where the bin widths are calculated by assuming a one-Parameter Laplace distribution for the errors, and fitting the parameter to the data, which amounts to calculating the mean. Another is to increase the number of bins m and restrict the search to the sorted coefficients ordered by decreasing absolute values. This illustrates the general feature of the *MDL* principle – that instead of a single criterion we can construct a sequence of them, one better than another, which process is restricted only by imagination. The reason for this is, of course, the common measure, the code length, while without it there is no general objective to provide guidance.

9.5 AR and ARMA Models

While modeling gaussian time series with AR models are instances of linear quadratic regression problems, their order estimation poses trouble for the stochastic complexity because the regressor matrix is determined by the parameters; and since the Fisher information is not constant, the integral of its square root is difficult to carry out. The same problem is also with the ARMA models, which have the additional difficulty of calculation of the maximum-likelihood parameters.

9.5.1 AR models

The likelihood density function for an AR model

$$y_t + \sum_{i=1}^{n} a_i y_{t-i} = e_t, \tag{9.66}$$

where we put $y_t = 0$ for $t < 1$, is given by

$$f(y^N; \theta) = \frac{1}{(2\pi\sigma^2)^{N/2}} e^{-\frac{1}{2\sigma^2} \sum_{t=1}^{N} (y_t + a_1 y_{t-1} + \ldots + a_n y_{t-n})^2}. \tag{9.67}$$

The maximized likelihood is $\dfrac{1}{(2\pi e\hat{\sigma}^2)^{N/2}}$, where $\hat{\sigma}^2$ is the minimized sum per symbol $\hat{\sigma}^2 = \dfrac{1}{N} \sum_{t=1}^{N} (y_t + \hat{a}_1 y_{t-1} + \ldots + \hat{a}_n y_{t-n})^2$. The NML criterion (5.40) then has the expression

$$L(y^N; n) = \frac{N}{2} \ln(2\pi e\hat{\sigma}^2) + \frac{n+1}{2} \ln \frac{N}{2\pi} + \ln \int_\Theta |\mathbf{J}(\theta)|^{1/2} d\theta + o(1), \tag{9.68}$$

where the Fisher information matrix is given by $J = \begin{bmatrix} \mathbf{R}_{zz} & 0 \\ 0 & 1/(2\sigma^4) \end{bmatrix}$, where

$$\mathbf{R}_{zz} = \begin{bmatrix} r_0 & r_1 & \cdots & r_{n-1} \\ r_1 & r_0 & \cdots & r_{n-2} \\ \vdots & \vdots & \ddots & \vdots \\ r_{n-1} & r_{n-2} & \cdots & r_0 \end{bmatrix},$$

and the $r_i = E[z_t z_{t-i}]$ denote the covariances of the process $z_t = y_t/\sigma$.

The evaluation of the integral in (9.68) can be done by the Monte Carlo technique. We refer to [20] for the actual implementation.

Example. Table 9.2 present, an evaluation of the NML, BIC, the predictive least-squares criterion PLS, and the bias-corrected KL Information $KICC$ criteria for estimating the order of AR models on a number of computer-generated data sequences.

We see that the NML criterion compares favorably with all the other criteria when the sample size is at least 50. In most of the cases BIC is ranked the second after the NML, and the results of $KICC$ do not improve when the sample size N is increased. For all criteria the performances decline for the larger values of the model order, which is clear because there is more to learn. Notice the moderate performances of the PLS criterion. This is to be expected since the PLS criterion is based on the estimates of the parameters, which are shaky for small amounts of data.

Table 9.2. The probability of correct estimation of the AR order (The best result for each sample size N is represented with bold font)

AR model order	Criterion	Sample size (N)			
		25	50	100	200
	NML	**0.99**	**0.99**	1.00	1.00
	BIC	0.93	0.95	0.97	0.98
$n = 1$	KICC	0.95	0.93	0.91	0.90
	PLS	0.89	0.92	0.95	0.97
	NML	0.72	**0.85**	**0.87**	**0.88**
	BIC	0.79	**0.85**	**0.87**	0.87
$n = 2$	KICC	**0.82**	0.83	0.80	0.78
	PLS	0.49	0.59	0.66	0.71
	NML	0.49	**0.74**	**0.83**	**0.84**
	BIC	**0.52**	0.71	0.78	0.79
$n = 3$	KICC	0.51	0.71	0.73	0.69
	PLS	0.26	0.39	0.47	0.53

9.5.2 ARMA models

Consider next the ARMA model:

$$y_t + \sum_{i=1}^{n} a_i y_{t-i} = e_t + \sum_{j=1}^{m} b_j e_{t-j} , \qquad (9.69)$$

where e_t is modeled by a zero-mean white gaussian noise of variance σ^2. The density function for the y_t's depends on how the initial values are related to the inputs e. A simple formula results if we put $y_i = e_i = 0$ for $i \leq 0$. Then the linear spaces spanned by y^t and e^t are the same. Let $\hat{y}_{t+1|t}$ be the orthogonal projection of y_{t+1} on the space spanned by y^t. We have the recursion

$$\hat{y}_{t+1|t} = \sum_{i=1}^{m} b_i(y_{t-i+1} - \hat{y}_{t-i+1|t-i}) - \sum_{i=1}^{n} a_i y_{t-i+1} , \qquad (9.70)$$

where $\hat{y}_{1|0} = 0$. With more general initial conditions the coefficients b_i in (9.70) will depend on t; see for instance [58].

The likelihood function of the model is

$$f(y^N; \theta, \sigma^2) = \frac{1}{(2\pi\sigma^2)^{N/2}} e^{-\frac{1}{2\sigma^2} \sum_{t=1}^{N}(y_t - \hat{y}_{t|t-1})^2} . \qquad (9.71)$$

The maximized likelihood is $\frac{1}{(2\pi e \hat{\sigma}^2)^{N/2}}$, where $\hat{\sigma}^2 = \min_{a_1,...,a_n,b_1,...b_m} \frac{1}{N} \sum_{t=1}^{N} (y_t - \hat{y}_{t|t-1})^2$. The NML criterion (5.40) is then given by

$$L(y^N; n, m) = \frac{N}{2} \ln(2\pi e \hat{\sigma}^2) + \frac{n+m+1}{2} \ln \frac{N}{2\pi} + \ln \int_\Theta |\mathbf{J}(\theta)|^{1/2} d\theta + o(1) \,.$$

$$(9.72)$$

This time the Fisher information matrix is more complicated than in the AR case. For a discussion of the calculation of varying generality depending on the nature of the roots we refer to [20].

Example. We calculate the structure of ARMA models for data generated by three different models. For each model, the true structure and the coefficients are given in Table 9.3, where we show the estimation results for 1,000 runs. In all experiments we have chosen the variance of the zero-mean white gaussian noise to be $\sigma^2 = 1$. There exist different methods for estimation of ARMA model parameters. We selected the one implemented in Matlab as armax function by Ljung, which is well described in his book [39].

Table 9.3. Results of model selection for the ARMA models in the example.[*]

ARMA model	Criterion	Sample size (N)				
		25	50	100	200	400
$n = 1, m = 1$	NML	700	**812**	**917**	**962**	**989**
$a_1 = -0.5$	BIC	638	776	894	957	983
$b_1 = 0.8$	KICC	**717**	740	758	745	756
$n = 2, m = 1$	NML	**626**	**821**	**960**	**991**	**994**
$a_1 = 0.64, a_2 = 0.7$	BIC	532	740	898	961	978
$b_1 = 0.8$	KICC	586	727	810	846	849
$n = 1, m = 1$	NML	851	**887**	**918**	**931**	**961**
$a_1 = 0.3$	BIC	766	804	856	903	942
$b_1 = 0.5$	KICC	**860**	764	654	614	577

[*]The counts indicate for 1,000 runs the number of times the structure of the model was correctly estimated by each criterion. The best result for each sample size N is represented with bold font.

9.6 Logit Regression

This work was done in [75].

The observations consist of n pairs $(y^n, \underline{x}^n) = (y_1, \underline{x}_1), \ldots, (y_n, \underline{x}_n)$, where the y_i are binary numbers and $\underline{x}_i \in \mathcal{R}^k$ vectors of quantized real numbers in such a way that there will be ℓ distinct vectors $\underline{x}_i \in \{\underline{b}_1, \ldots, \underline{b}_\ell\}$ for $\ell < n$. Let the number of occurrences of $\underline{b}_i$ in the string $\underline{x}_1, \ldots, \underline{x}_n$ be n_i, and let $y_i = 1$ be $m_i(y^n) = m_i$ times of these occurrences. We can actually take $n_i > 1$, for if $n_i = 1$ for some i, it could be removed since we would have only one occurrence of the value y_i, and the conditional m_i/n_i is either 0 or

1, which would add nothing to the statistical relationship $\{y_j|\underline{x}_j\}$, defined by the conditionals

$$P(Y_i = 1|\underline{x}_i; \underline{\beta}) = \frac{e^{\underline{\beta}^T \underline{x}_i}}{1 + e^{\underline{\beta}^T \underline{x}_i}}$$

$$P(Y_i = 0|\underline{x}_i; \underline{\beta}) = \frac{1}{1 + e^{\underline{\beta}^T \underline{x}_i}},$$

where $\underline{\beta} = \beta_1, \ldots, \beta_k$ (see [13, 30]). Notice, however, that if we add the Bernoulli models to the family by ignoring all the $\underline{x}_i$'s, then, of course, we should include all occurrences of y_i.

We extend $P(Y_i = 1|\underline{x}_i; \underline{\beta})$ to the sequence $(y^n, \underline{x}^n)$ by independence so that

$$P(Y^n|\underline{x}^n; \underline{\beta}) = \frac{\prod_{i=1}^{\ell} e^{m_i \underline{\beta}^T \underline{b}_i}}{\prod_{i=1}^{n}(1 + e^{\underline{\beta}^T \underline{x}_i})} = \frac{e^{\sum_{i=1}^{\ell} m_i \underline{\beta}^T \underline{b}_i}}{\prod_{i=1}^{n}(1 + e^{\underline{\beta}^T \underline{x}_i})} = \frac{e^{\underline{\beta}^T \underline{t}}}{\prod_{i=1}^{\ell}(1 + e^{\underline{\beta}^T \underline{b}_i})^{n_i}}, \tag{9.73}$$

where $\underline{t} = \sum_{i=1}^{\ell} m_i \underline{b}_i$ is a sufficient statistic. Notice again that if some $n_i = 1$, the pair $(y_i, \underline{x}_i)$ could be dropped and n reduced. Hence, for the full advantage of the logit models all n_i should exceed unity.

The ML estimates of parameters

The following maximum-likelihood estimates are well known, but we write them in terms of the notations n_i and m_i, which simplify the subsequent algorithm. We have

$$\log P(y^n|\underline{x}^n; \underline{\beta}) = \underline{\beta}^T \underline{t} - \sum_{i=1}^{\ell} n_i \log(1 + e^{\underline{\beta}^T \underline{b}_i}) \tag{9.74}$$

and

$$\frac{\log P(Y^n|\underline{x}^n; \underline{\beta})}{d\beta} = \underline{t} - \sum_{i=1}^{\ell} n_i \underline{b}_i \frac{e^{\underline{\beta}^T \underline{b}_i}}{(1 + e^{\underline{\beta}^T \underline{b}_i})}. \tag{9.75}$$

The ML parameter vector $\hat{\underline{\beta}}$ satisfies

$$\underline{t} = \sum_{i=1}^{\ell} n_i \underline{b}_i \frac{e^{\hat{\underline{\beta}}^T \underline{b}_i}}{(1 + e^{\hat{\underline{\beta}}^T \underline{b}_i})}, \tag{9.76}$$

from which $\hat{\underline{\beta}}(y^n) = \hat{\underline{\beta}}_{\underline{t}}$ can be solved by numerical means, or equivalently from

$$\sum_{i=1}^{\ell} n_i \underline{b}_i \left(\frac{m_i}{n_i} - \frac{e^{\hat{\underline{\beta}}^T \underline{b}_i}}{1 + e^{\hat{\underline{\beta}}^T \underline{b}_i}} \right) = 0. \tag{9.77}$$

The normalized maximum-likelihood model

The normalized maximum likelihood model for each collection of the retained regressor variables or the corresponding parameters, which define the different model classes, is as follows:

$$\hat{P}(y^n|\underline{x}^n) = \frac{P(y^n|\underline{x}^n;\hat{\underline{\beta}})}{C(X)}, \tag{9.78}$$

where

$$P(y^n|\underline{x}^n;\hat{\beta}) = \frac{e^{\hat{\beta}^T t}}{\prod_{i=1}^{\ell}(1 + e^{\hat{\beta}^T \underline{b}_i})^{n_i}} = P(y^n|\underline{t};\hat{\underline{\beta}}_{\underline{t}}). \tag{9.79}$$

The normalizing coefficient can be written as

$$C(X) = \sum_{y^n} P(y^n|\underline{x}^n;\hat{\beta}) = \sum_{\underline{t}\in\Omega} c_n(\underline{t}, X) P(y^n|\underline{t};\hat{\underline{\beta}}_{\underline{t}})$$

$$= \sum_{\underline{t}\in\Omega} c_n(\underline{t}, X) \frac{e^{\hat{\underline{\beta}}_{\underline{t}}^T \underline{t}}}{\prod_{i=1}^{\ell}(1 + e^{\hat{\underline{\beta}}_{\underline{t}}^T \underline{b}_i})^{n_i}}, \tag{9.80}$$

where $c_n(\underline{t}, X)$ denotes the number of strings y^n such that $\sum_{i=1}^{\ell} m_i(y^n)\underline{b}_i = \underline{t}$. The sum is taken over all $\underline{t}$ such that

$$\underline{t} = \sum_{i=1}^{\ell} m_i\underline{b}_i$$

$$m_i \leq n_i, \ i = 1, \ldots, \ell$$

$$\sum_{j=1}^{\ell} n_i = n.$$

An algorithm, worked out by Ioan Tabus, quickly calculates the normalizing coefficient for data-set sizes that cover the usual cases. It takes advantage of the fact that there are repeated occurrences of the regressor vectors $\underline{b}_i$. An outline of the algorithm follows.

There are $\prod_{i=1}^{\ell}(n_i + 1)$ possibilities for the numbers $m_1, m_2, \ldots, m_\ell$, some of which give the same value for the sufficient statistic $\underline{t} = \sum_{i=1}^{\ell} m_i\underline{b}_i$.

We compute recursively in L, $L = 1, 2, \ldots, \ell$, the set $\Delta = \{\underline{t}|\underline{t} = \sum_{i=1}^{\ell} m_i\underline{b}_i\}$ of achievable sufficient statistic vectors and the computation of the counts of each. Denote by $\underline{t}_L = \sum_{i=1}^{L} m_i\underline{b}_i$ one partial sum up to step L, and $\Delta_L = \{\underline{t}_L : \underline{t}_L = \sum_{i=1}^{L} m_i\underline{b}_i,\}$ where $m_i \in \{0, 1, \ldots, n_i\}$.

The main recursion for constructing the sufficient statistics vectors is $\underline{t}_L^+ = \underline{t}_{L-1} + j\underline{b}_L$, where $j \in \{0, 1, \ldots, n_L\}$. The sketch of the algorithm is as follows:

0. Initialize $\Delta_1 = \Delta$ with the vectors $j\underline{b}_1$, $j = 1 : n_1$
1. For $L = 2 : \ell$
    1.1 For $j = 1 : n_L$
        1.1.1 For $i = 1 : |\Delta_{L-1}|$
            1.1.1.1 $\underline{t}_L^+ = \underline{t}_{L-1}(i) + j\underline{b}_L$
            1.1.1.2 For $k = 1 : |\Delta|$
                If $\underline{t}_L^+$ is identical to $\underline{t}(k)$, the kth vector in Δ,
                update counts: $c(\underline{t}_L(k)) = c(\underline{t}_L(k)) + \binom{n_L}{j} c(\underline{t}_{L-1}(i)))$
            1.1.1.3 If $\underline{t}_L^+ \notin \Delta_{L-1}$, append $\underline{t}_L^+$ to Δ and set $c(\underline{t}_L^+) = \binom{n_L}{j} c(\underline{t}_{L-1}(i)))$
        1.1.2 Set $\Delta_L = \Delta$

The so-obtained *NML* criterion was applied to the data in Table 3 in [30]. For $k = 3$ there are the $\ell = 8$ distinct (column) regressor vectors $000, 001, \ldots, 111$, repeated $3, 2, 4, 5, 1, 5, 8, 17$ times, respectively, at which $y = 1$ occurs $3, 2, 44, 5, 1, 3, 5, 6$ times, respectively.

We give in Table 9.4 the *NML* code length and the logarithm of the normalizing coefficient in columns 2 and 3, respectively, for all the eight subsets of the retained regressor variables listed in column 1. For instance, the index set $\gamma = 110$ means that the first two variables b_1 and b_2 are retained while the third is ignored, and for $k = 2$ there are occurrences of the $\ell = 4$ distinct values of the regressor variables, and the parameters fitted are β_1 and β_2. The *NML* model for the class defined for this subset is seen to assign the largest probability to the observed data and hence provides the best explanation of the statistical dependency $Y|X$. We also see that the first variable is the most important, and the third provides little additional information to the first two. We clearly see that the *NML* models give much more information than the traditional separate testing procedures. In fact, the traditional hypothesis testing procedures consist of "accept–reject" rules on arbitrarily set levels for the null hypotheses of type $\beta_i = \beta_j = \ldots = 0$ without giving quantitative comparative importance of the various variables. Moreover, as discussed in the subsection Hypothesis Testing, traditional testing procedures do not take

Table 9.4. Logit regression (Optimum in boldface)

γ	$\mathbf{L_{NML}}$	$\log_2 \mathbf{C(X)}$
000	46.91	3.20
001	45.21	5.48
010	44.88	5.41
100	41.18	5.26
011	43.84	7.98
101	41.31	7.76
110	**40.49**	7.72
111	41.34	10.47

into account the effect of the possible errors in the fitted models nor the fact that there are different numbers of parameters.

Historical Notes

The applications reported here are only those I have either worked out myself or have participated in. Since the introduction of the *NML* model there have been a number of successful applications of it to DNA-related work because the normalizing coefficient can be computed exactly (see [74] and the references therein).

References

1. Akaike, H. (1973) "Information theory as an extension of the maximum likelihood principle," pages 267–281 in B.N. Petrov and F. Csaki (eds): *Second International Symposium on Information Theory*. Akademiai Kiado, Budapest.
2. Atteson, K. (1999) "The Asymptotic Redundancy of Bayes Rules for Markov Chains," *IEEE Trans. Information Theory*, Vol. **IT-45**, No. 6, September 1999.
3. Balasubramanian, V. (1996) "Statistical Inference, Occam's Razor and Statistical Mechanics on the Space of Probability Distributions," *Neural Computation*, **9**, No. 2, 349–268, 1997 http://arxiv.org/list/nlin/9601.
4. Balasubramanian, V. (2005) "MDL, Bayesian Inference, and Geometry of the Space of Probability Distributions," chapter 3 in *Advances in Minimum Description Length: Theory and Applications*, Peter D. Grünwald, In Jae Myung and Mark A. Pitt, eds. A Bradford Book, The MIT Press, Cambridge, Massachusetts.
5. Barron, A.R. (1985) "Logically Smooth Density Estimation," Ph.D. dissertation, Dept. of EE, Stanford University.
6. Boltzmann, L. (1895) *Vorlesungen uber Gastheorie*, 1er Theil, Leibzig, Barth.
7. Barron, A.R., Rissanen, J., and Yu, B. (1998) "The MDL Principle in Modeling and Coding," special issue of *IEEE Trans. Information Theory* to commemorate 50 years of information theory, Vol. **IT-44**, No. 6, October 1998, pp 2743–2760.
8. Boltzmann, L. (1895) *Vorlesungen uber Gastheorie*, 1er Theil, Leibzig, Barth.
9. Chaitin, G.J. (1969) "On the Length of Programs for Computing Finite Binary Sequences: Statistical Considerations," *JACM*, **16**, 145–159.
10. Clarke, B.C. and Barron, A.R. (1990) "Information-Theoretic Asymptotics of Bayes Methods," *IEEE Trans. Information Theory*, Vol. **IT-36**, pp 453–471.
11. Cover, T. (1973) "Enumerative Source Encoding," *IEEE Trans. Information Theory*, Vol. **IT-19**, pp 73–77.
12. Cover, T. and Thomas, J. (1991) *Elements of Information Theory*, John Wiley and Sons, Inc., New York, 542 pages.
13. Cox, D.R. (1970) *The Analysis of Binary Data*, London: Methuen.
14. Davis, M.H.A. and Hemerly, E.M. (1990) "Order Determination and Adaptive Control of ARX Models Using the PLS Criterion", *Proceedings of the Fourth Bad Honnef Conference on Stochastic Differential Systems*. Lecture Notes in Control and Information Sci. (N. Christopeit, ed.) Springer, New York.

15. Dawid, A.P. (1984) "Present Position and Potential Developments: Some Personal Views, Statistical Theory, The Prequential Approach," *J. Royal Stat. Soc. A*, Vol. **147**, Part 2, 278–292.

16. Drmota, M., Reznik, Y., Savari, S., and Szpankowski, W. (2006) "Precise Asymptotic Analysis of the Tunstall Code," *IEEE Intern. Symposium on Information Theory*, Seattle.

17. Donoho, D.L. and Johnstone, I.M. (1994) "Ideal Spatial Adaptation by Wavelet Shrinkage," *Biometrika*, **81**, 425–455.

18. Elias, P. (1975) "Universal Codeword Sets and Representation of Integers," *IEEE Trans. Information Theory*, Vol. **IT-21**, 194–203.

19. van Erven, T. (2006) "The Momentum Problem in MDL and Bayesian Prediction," Master of Science thesis, Universiteit van Amsterdam.

20. Giurcăneanu, C.D. and Rissanen, J. (2006) "Estimation of AR and ARMA Models by Stochastic Complexity," in the procedings of CZ Wei memorial workshop, December 2005, Academia Sinica, *IMS Lecture Notes-Monograph Series*.

21. Gradshteyn, I.S. and Ryzhik, I.M. (1980) *Table of Integrals, Series and Products*, Academic Press, New York, 1160 pages.

22. Grünwald, P.D. (1998) *The Minimum Description Length Principle and Reasoning Under Uncertainty*, Ph.D. thesis, Institute for Logic, Language and Computation, Universiteit van Amsterdam, 296 pages.

23. Grünwald, P.D. (2005), 'Tutorial on MDL', pp 1–57 in *Advances in Minimum Description Length: Theory and Applications*, Peter D. Grünwald, In Jae Myung and Mark A. Pitt, eds. A Bradford Book, The MIT Press, Cambridge, Massachusetts.

24. Grünwald, P.D. (2007) The Minimum Description Length Principle, MIT Press.

25. Hall, P. and Hannan, E.J. (1988) "On Stochastic Complexity and Nonparametric Density Estimation," *Biometrika*, vol. **75**, pp 705–714, December 1988.

26. Hamming, R.W. (1980) *Coding and Information Theory*, Prentice-Hall, Englewood Cliffs, N.J., 239 pages.

27. Hansen, M.H. and Yu, B. (2001) "Model Selection and the Principle of Minimum Description Length," *Journal of American Statistical Association*, **96**(454), 746–774.

28. Hansen, M.H. and Yu, B. (2000) "Wavelet Thresholding via MDL for Natural Images," *IEEE Trans. Information Theory* (Special Issue on Information Theoretic Imaging), **46**, 1778–1788.

29. Hartley, R.V. (1928) "Transmission of Information," Bell System Technical Journal, **7**, 535–563.

30. Hirji, K.F., Mehta, C.R., and Patel, N.R. (1987) "Computing Distributions for Exact Logistic Regression," *Journal of the American Statistical Association*, Vol. **82**, No. 400, pp 1110–1117, Dec. 1987.

31. Hoel, A.E. and Kennard, R.W., "Ridge Regression: Biased Estimation for Nonorthogonal Problems," *Technometrics*, **12**, 55–68.

32. James, W. and Stein, C. (1961) "Estimation with Quadratic Loss," *Proc. 4th Berkeley Symp.* **1**, 363–379.

33. Jeffreys, H. (1961) *Theory of Probability*, Clarendon Press, Oxford, 447 pages, (third edition).

34. Kolmogorov, A.N. (1965) "Three Approaches to the Quantitative Definition of Information," *Problems of Information Transmission* **1**, 4–7.

35. Kolmogorov, A.N. (1968) "Logicl Basis for Information Theory and Probability Theory," *IEEE Trans. Information Theory*, Vol. **IT-14**, 662–664.

36. Kumar, V., Heikkonen, J., Rissanen, J., and Kaski, K. (2005) "MDL Denoising with Histogram Models," *IEEE Transactions on Signal Processing*, **54**, Nr 8, 2922–2928, August 2006.

37. Lempel, A. and Ziv, J., "Compression of Individual Sequences via Variable Rate Coding," *IEEE Trans. Information Theory*, Vol. **IT-24**, 530–536, September 1978.

38. Li, M. and Vitanyi, P.M.B. (1997) *An Introduction to Kolmogorov Complexity and its Applications*, Springer-Verlag, New York, Second Edition, (xx+637 pages).

39. Ljung, L. (1999). *System Identification : Theory for the User*, 2nd ed. Prentice Hall, Upper Saddle River, N.J.

40. Leung-Yan-Cheong, S.K. and Cover, T. (1978) "Some Equivalences Between Shannon Entropy and Kolmogorov Complexity," *IEEE Trans. Information Theory*, Vol. **IT-24**, 331–338.

41. Marvin L. Minsky (1967) Computation: Finite and Infinite Machines, Prentice-Hall, Inc., Englewood Cliffs, N.J., 317.

42. Merhav, N. and Feder, M. (1995) "A Strong Version of the Redundancy-Capacity Theorem of Universal Coding," *IEEE Trans. Information Theory*, Vol. **IT-41**, No. 3, pp 714–722, May 1995.

43. Nohre, R. (1994) *Some Topics in Descriptive Complexity*, Ph.D. Thesis, Linkoping University, Linkoping, Sweden.

44. Pasco, R. (1976) *Source Coding Algorithms for Fast Data Compression*, Ph.D. thesis, Stanford University.

45. Department of Mathematics, Statistics and Philosophy, Liski, E.P., isotalo, J., Niemel, J., Puntanen, S., and Styan, G.P.H. (eds), Report A 368, ISBN 978-951-44-6620-5, pp 159–172.

46. Qian, Q. and Künsch, H.R. (1998) "Some notes on Rissanen's Stochastic Complexity, *IEEE Trans. Information Theory*, Vol. **IT-44**, Nr. 2, March 1998.

47. Rao, C.R. (1975) "Simultaneous Estimation of Parameters in Different Linear Models and Applications," *Biometrics*, **31**, 545–554.

48. Rao, C.R. (1981) "Prediction of Future Observations in Polynomial Growth Curve Models," *Proc. Indian Stat. Inst. Golden Jubilee Int. Conf. on Statistics: Applications and New Directions*, 512–520.

49. Rissanen, J. and Langdon, G.G. Jr, "Universal Modeling and Coding," *IEEE Trans. Information Theory*, Vol. **IT-27**, Nr. 1, 12–23.

50. Rissanen, J. (1976) "Generalized Kraft Inequality and Arithmetic Coding," IBM J. Res. Dev. **20**, Nr 3, 198–203.

51. Rissanen, J. (1978) "Modeling by Shortest Data Description," *Automatica*, Vol. **14**, pp 465–471.

52. Rissanen, J. (1976) "Arithmetic Codings as Number Representations," *Acta Polytechnica Scandinavica*, Ma 31, Helsinki 1979, pp 44–51.

53. Rissanen, J. (1983) "A Universal Data Compression System," *IEEE Trans. Information Theory*, Vol. **IT-29**, Nr. 5, 656–664.

54. Rissanen, J. (1983) "A Universal Prior for Integers and Estimation by Minimum Description Length," *Annals of Statistics*, Vol. **11**, No. 2, 416–431.

55. Rissanen, J. (1984) "Universal Coding, Information, Prediction, and Estimation," *IEEE Trans. Information Theory*, Vol. **IT-30**, Nr. 4, 629–636.

56. Rissanen, J. (1986) "Stochastic Complexity and Modeling," *Annals of Statistics*, Vol. **14**, 1080–1100.

138   References

57. Rissanen, J. (1986) "A Predictive Least Squares Principle", IMA J. Math. Control Inform. **3**, 211-222

58. Rissanen, J. (1989) *Stochastic Complexity in Statistical Inquiry*, World Scientific Publ. Co., Suite 1B, 1060 Main Street, River Edge, N.J. (175 pages).

59. Rissanen, J. (1992) "Noise Separation and MDL Modeling of Chaotic Processes," pp 317–330 in P. Grassberger and J.-P. Nadal (eds): *From Statistical Physics to Statistical Inference and Back*, Kluwer Academic Publishers, Netherlands.

60. Rissanen, J. (1996) "Fisher Information and Stochastic Complexity," *IEEE Trans. Information Theory*, Vol. **IT-42**, No. 1, pp 40–47.

61. Rissanen, J. (2000) "MDL Denoising," *IEEE Trans. on Information Theory*, Vol. **IT-46**, Nr. 7, November 2000.

62. Rissanen, J. (2001) "Strong Optimality of the Normalized ML Models as Universal Codes and Information in Data," *IEEE Trans. Information Theory*, Vol. **IT-47**, Nr. 5, July 2001.

63. Rissanen, J. (2006) "Distinguishable Distributions" (submitted for publication).

64. Rissanen, J., Speed, T.P., and Yu, B. (1992) "Density estimation by stochastic complexity," *IEEE Trans. Information Theory*, Vol. **IT-38**, Nr. 2, March 1992.

65. Rissanen, J. and Yu, B. (2000) "Coding and Compression: A Happy Union of Theory and Practice," JASA, (Invited Year 2000 Commemorative Vignette on Engineering and Physical Sciences), **95**, 986–988.

66. Rissanen, J. and Tabus, I. (2005) "Kolmogorov Structure Function in MDL Theory and Lossy Data Compression," chapter 10 in Jae Myung and Mark A. Pitt (eds): *Advances in Minimum Description Length: Theory and Applications*, Peter D. Grünwald, A Bradford Book, The MIT Press, Cambridge, Massachusetts.

67. Roos, T., Myllymaki, P., and Rissanen, J., "MDL Denoising Revisited" (submitted for publication).

68. Shannon, C.E. (1948) "A Mathematical Theory of Communication," Bell System Technical Journal, **27**, 379–423, 623–656.

69. Shtarkov, Yu. M. (1987) "Universal Sequential Coding of Single Messages," translated from *Problems of Information Transmission*, Vol. 23, No. 3, 3–17, July–September 1987.

70. Solomonoff, R.J. (1960) "A Preliminary Report on a General Theory of Inductive Inference," Report ZTB-135, Zator Co., Cambridge, Mass., November 1960.

71. Solomonoff, R.J. (1964) "A Formal Theory of Inductive Inference," Part I, *Information and Control*, **7**, 1–22; Part II, *Information and Control*, **7**, 224–254.

72. Szpankowski, W. (1998) "On Asymptotics of Certain Recurrences Arising in Universal Coding," *Problems of Information Transmission*, **34**, No. 2, 142–146.

73. Tabus, I. and Rissanen, J. (2002) "Asymptotics of Greedy Algorithms for Variable-to-Fixed Length Coding of Markov Sources," *IEEE Trans. Information Theory*, Vol. **IT-48**, pp 2022–2035, July 2002.

74. Tabus I., Rissanen, J., and Astola, J. (2006) "Classification and Feature Gene Selection Using the Normalized Maximum Likelihood Model for Discrete Regressions," *Signal Processing*, Special issue on genomic signal processing; also available at http://sigwww.cs.tut.fi/ tabus/GSP.pdf

75. See reference 41 except for page numbers pp. 295–300.

76. Takeuchi, Jun-ichi and Barron, A.R. (1998) "Robustly Minimax Codes for Universal Data Compression," *The 21st Symposium on Information Theory and Its Applications*, Gifu, Japan, December 2–5.

77. Takeuchi, Jun-ichi, Kawabata, Tsutomu, and Barron, A.R. (2001) "Properties of Jeffreys' Mixture for Markov Sources," *2001 Workshop on Information-Based Induction Sciences*, Tokyo, Japan, July 30–August 1, 2001.

78. Takimoto, E. and Warmuth, M. (2000) "The Last-Step Minimax Algorithm", *Proceedings of the 11'th International Conference on Algorithmic Learning Theory*

79. Vereshchagin, N.K. and Vitanyi, P.M.B. (2004) "Kolmogorov's Structure Functions and Model Selection," *IEEE Trans. Information Theory*, Vol. **IT-50**, Nr. 12, 3265–3290.

80. Vovk, V.G. (1990) "Aggregating Strategies," Proceedings of 3rd Annual Workshop on Computational Learning Theory, Morgan Kauffman, pp 371–386.

81. Wallace, C.S. and Boulton, D.M. (1968) "An Information Measure for Classification," *Computing Journal*, **11**, 2, 185–195.

82. Wei, C.Z. (1992), "On Predictive Least Squares Principles", *The Annals of Statistics*, Vol. **20**, No. 1, pp 1-42, March 1992.

83. Weinberger, M.J., Rissanen, J., and Feder, M. (1995) "A Universal Finite Memory Source," *IEEE Trans. on Information Theory*, Vol. **IT-41**, No. 3, pp 643–652, May 1995.

84. White, H. (1994) *Estimation, Inference, and Specification Analysis*, Cambridge University Press, Cambridge, United Kingdom, 349 pages.

85. Wiener, N. (1961) *Cybernetics* (First edition, 1948), Second edition (revisions and two additional chapters), The MIT Press and Wiley, New York.

86. Wyner, A.D. (1972) "An Upper Bound on Entropy Series," *Information and Control*, **20**, 176–181.

87. Yamanishi, K. (1998) "A Decision-Theoretic Extension of Stochastic Complexity and Its Application to Learning," *IEEE Trans. on Information Theory*, Vol. **IT-44**, No. 4, July 1998.

Index

All of Statistics

L. Wasserman

This book is for people who want to learn probability and statistics quickly. It brings together many of the main ideas in modern statistics in one place. The book is suitable for students and researchers in statistics, computer science, data mining and machine learning. This book covers a much wider range of topics than a typical introductory text on mathematical statistics. It includes modern topics like nonparametric curve estimation, bootstrapping and classification, topics that are usually relegated to follow-up courses.

2004. 442 p. (Springer Texts in Statistics) Hardcover ISBN 978-0-387-40272-7

Statistical and Inductive Inference by Minimum Message Length

C.S. Wallace

This book gives a sound introduction to the Minimum Message Length Principle and its applications, provides the theoretical arguments for the adoption of the principle, and shows the development of certain approximations that assist its practical application. MML appears also to provide both a normative and a descriptive basis for inductive reasoning generally, and scientific induction in particular. The book describes this basis and aims to show its relevance to the Philosophy of Science.

2005. 432 p. (Information Science and Statistics) Hardcover
ISBN 978-0-387-23795-4

Pattern Recognition and Machine Learning

Christopher M. Bishop

The dramatic growth in practical applications for machine learning over the last ten years has been accompanied by many important developments in the underlying algorithms and techniques. This completely new textbook reflects these recent developments while providing a comprehensive introduction to the fields of pattern recognition and machine learning. It is aimed at advanced undergraduates or first-year PhD students, as well as researchers and practitioners. No previous knowledge of pattern recognition or machine learning concepts is assumed.

2006. 702 p. (Information Science and Statistics) Hardcover
ISBN 978-0-387-31073-2

Easy Ways to Order▶ Call: Toll-Free 1-800-SPRINGER · E-mail: orders-ny@springer.com · Write: Springer, Dept. S8113, PO Box 2485, Secaucus, NJ 07096-2485 · Visit: Your local scientific bookstore or urge your librarian to order.